Consanguinity, Inbreeding, and
Genetic Drift in Italy

MONOGRAPHS IN POPULATION BIOLOGY
EDITED BY SIMON A. LEVIN AND HENRY S. HORN

Complete series list follows Index.

# Consanguinity, Inbreeding, and Genetic Drift in Italy

LUIGI LUCA CAVALLI-SFORZA,

ANTONIO MORONI, AND

GIANNA ZEI

PRINCETON UNIVERSITY PRESS

Princeton and Oxford

Copyright © 2004 by Princeton University Press
Published by Princeton University Press, 41 William Street,
Princeton, New Jersey 08540
In the United Kingdom: Princeton University Press, 3 Market Place,
Woodstock, Oxfordshire OX20 1SY

All Rights Reserved

**Library of Congress Cataloging-in-Publication Data**

Cavalli-Sforza, L. L. (Luigi Luca), 1922–
Consanguinity, inbreeding, and genetic drift in Italy / Luigi Luca Cavalli-Sforza,
Antonio Moroni and Gianna Zei.
p. cm. — (Monographs in population biology ; 39)
Includes bibliographical references and index.
ISBN 0-691-08991-4 — ISBN 0-691-08992-2 (pbk.)
1. Consanguinity—Italy. 2. Inbreeding—Italy. 3. Human population genetics—
Italy. 4. Human genetics–Variation. I. Moroni, Antonio. II. Zei, Gianna.
III. Title. IV. Series.
GN480.25.C38 2004
304.6'0945—dc21        2003053599

British Library Cataloging-in-Publication Data is available

This book has been composed in Times Roman

Printed on acid-free paper. ∞

www.pupress.princeton.edu

Printed in the United States of America

10  9  8  7  6  5  4  3  2  1

# Contents

# Acknowledgments

This book summarizes research started in 1952 through the collaboration of the first two authors; the third joined in 1959. During this time many people have contributed and their help is very gratefully acknowledged. In particular, this study was made possible by the support of Cardinal Giuseppe Casoria and the cooperation of the Directors of Episcopal Archives of the Piacenza, Parma, Reggio Emilia, Bressanone, Sardinia, and Sicily Dioceses.

Among those who are responsible for major contributions to these investigations, it is a pleasure to mention Italo Barrai, now professor emeritus of genetics at the University of Ferrara, who was a major collaborator in research developed in chapters 3, 5, and 6, and to a lesser extent in other chapters; and Franco Conterio, now professor emeritus of anthropology at the University of Parma, who was responsible for the entire effort of collecting blood samples of the Parma valley, and for several investigations of inbreeding effects (chapters 5 and 8). Blood group tests were carried out by Mario Turri at the Istituto Sieroterapico Milanese Serafino Belfanti, Milan.

At Parma University, Professor Bruno Schreiber was of major help and support from 1951 until his retirement. Also at Parma University, Paolo Menozzi, professor of ecology in the Ecology Department, has been extremely helpful in various stages of this work.

Marcella Devoto, now chief of the Laboratory of Genetic Epidemiology of the Alfred duPont Hospital di Willmington, was especially kind in providing a preliminary version of chapter 8.

Among students working for their thesis in Natural or Biological Sciences and research assistants at Parma or Pavia Universities, Carla Zalaffi, Marisa Mainardi, Lamberto Soliani and Paola Astolfi made valuable contributions.

Enzo Siri, Walter Anghinetti, Romano Zanni, and Aldo Anelli of Parma University did much of data collection from Catholic documents; Enzo Siri also provided valuable help in the advanced stages of this work. Father Giuseppe Bracco directed a team of students who transferred to computer tapes about 540,000 consanguinity dispensations available in the Vatican archives.

A special thanks is addressed to Ornella Fiorani and Antonella Lisa of the Istituto di Genetica Molecolare of Pavia for the tremendous work they carried out for many years on surname studies, with competence, ability, and patience in the collection of data, computer programming, statistical analyses, and production of geographic maps.

We thank the referees of this book for useful remarks and, in particular, for their suggestion to add to chapter 11 our views on important problems that are still unresolved and possible approaches to working on them. It was our pleasure to include this information.

Financial support for this research has come, in the initial period (1954–1958) from the Rockefeller Foundation; from the U.S. Atomic Energy Commission (today's Department of Energy) in the period 1957–1971; from the Comitato Nazionale di Ricerca Nucleare in the period 1957–1962, and from the Consiglio Nazionale delle Ricerche beginning in 1962, through the Laboratorio and later Istituto di Genetica Biochimica and Evoluzionistica of Pavia, today Istituto di Genetica Molecolare del CNR.

# History of This Investigation
# and Structure of This Book

## 1.1 INTRODUCTION

The study of inbreeding, the consequence of the mating of relatives, has an important place in genetics. The similarity of the paternal and maternal contributions caused by the mating of relatives leads to increased genetic homogeneity of inbred individuals. A table of the expected effects of inbreeding in successive generations of selfing (crossing with self), the closest mating possible, which often occurs spontaneously in many plants, appears in Mendel's article, the founding paper of genetics. Thus, Mendel was also the first population geneticist.

Human societies are unique in keeping records of their own ancestry, sometimes, though very rarely, for thousand of years or more. Some breeders of domestic animals, however, have kept records of their animals' ancestry for even longer periods, measured in terms of numbers of generations.

The genetic effect of inbreeding can be estimated by the increase in average homozygosity over that expected by random mating. *Homozygosity* is the average percent frequency of homozygotes—individuals receiving the same form of a gene from both parents. The complement to 100% of homozygosity is *heterozygosity*.

Matings of close relatives (often called consanguineous matings), if repeated for many generations, increase homozygosity to a point of seriously decreasing fertility and individual survival, making prolonged inbreeding incompatible with continuation of life. To obtain stocks of high genetic homogeneity, animal breeders often make recourse to systematic parent–child or brother–sister matings for a number of generations (twenty or more). In the process they lose many inbred lines because of loss of fertility or survival, and at best end with stocks of weak or delicate constitution (inbreeding depression). There are examples of repeated brother–sister marriages in ancient Egyptian and Persian dynasties; but, for reasons probably not of genetic nature, no dynasty lasted for periods long enough to expect serious inbreeding depression. In any case we have no records from which to observe it. But apart from these examples of encouragement of brother–sister marriages, which was still popular in the Egypt of Cleopatra's time, unions between brother and sister and parent and child are condemned and avoided in practically all human societies. The term "incest" refers to these tabooed unions. They occur, though rarely. Children of incest are very rare, of the order of 1/10,000 in an estimate in Michigan (Adams and Neel 1967).

The dangers of close inbreeding must have become known to early humans fairly soon, since practically every society has rules that tend to avoid close consanguineous marriage, and sometimes extend prohibition to remote consanguinities. This is especially true of very small communities, like those of Eskimos living in extreme northern latitudes (Sutter and Tabah 1956), which are under greater risk of reaching high levels of inbreeding. In one remote, very small, and highly isolated Greenland Eskimo community, no marriage closer than third cousins was found in genealogies from living individuals.

Animals also tend to avoid close inbreeding, by social customs that seem to have been favored by natural selection in response to inbreeding depression. In most Primates, social groups are fairly permanent, but young males reaching puberty tend to leave the group into which they were born and join other groups. This custom is clearly effective in limiting close consanguineous matings. Chimpanzees are the only exception, as here it is the females that leave the group. Among humans one likewise observes a greater tendency of females to leave their birthplace at marriage. Wives tend to move to their husbands' residence in 70% of traditional societies (Murdock 1967). This custom has important genetic consequences: mitochondrial DNA (mtDNA), which is transmitted by the maternal line, should show less geographic clustering than Y chromosomes, which are transmitted by the male line. This expectation was confirmed by observation (Seielstad et al. 1998). In fact, the Y chromosome shows greater genetic variation between populations than nonsexual chromosomes and mitochondrial DNA.

In addition, at least in chimpanzees and other mammals, some tendency to avoid brother–sister or parent–child mating is observed. Among humans, the social custom of avoiding marriage of close relatives is paralleled by a similar constitutional safeguard against brother–sister mating, in the form of the so-called Westergaard effect: a tendency to avoid sexual contacts between brother and sister, or, in general, children who have been brought up together. Puberty seems to be the dividing line between social contacts that are unfavorable (before puberty) or favorable (after it) to interest in establishing sexual relations between individuals of opposite sex. Research has shown that children brought up in the same kibbutz, where they were usually raised together, marry very rarely, if ever. An old Chinese custom, which survived in Taiwan until recently, is the adoption of a young girl by a family in which a son was born, so that this girl becomes his future wife. These so-called "minor" marriages have been shown to be, on average, less fertile and less long-lasting than ordinary marriages with girls not brought up in the family (Wolf 1980).

There are, however, social exceptions to the rule of avoidance of close consanguineous marriages, less close than brother–sister. In certain social groups such marriages may be much more popular than would be expected by chance, undoubtedly because of a social preference. In West and South Asia two types of consanguineous marriages are especially common: uncle–niece marriages comprise up to 20% of all marriages in several north Indian tribes, and first-cousin marriages reach 50% or more in many Middle Eastern ethnic groups (Arabs, North Africans, and some Jewish groups). First-cousin marriages are or were high in Japan, especially at a time when marriages were mostly arranged by parents. High consanguinity customs spread around with the people who developed them. Perhaps as a remote consequence of Arab domination in Sicily and southern Italy in the eighth to the eleventh centuries, the frequency of uncle–niece and first-cousin matings became high in these regions and is currently especially high in Sicily. These relatively moderate degrees of inbreeding do not seem to have had a truly damaging effect. They may have contributed to lowering the current frequency of lethal and semilethal genetic diseases, at least in Japanese populations (Cavalli-Sforza and Bodmer 1971, 1999).

By definition, recessive genes are those that are manifested only in homozygotes. In inbred families increased homozygosity is expected, leading to a higher probability of observing recessive inheritance. The study of consanguineous marriages, therefore, has merit for the detection of recessive genes and for the study of their frequency.

In this chapter we summarize the salient points of the history of this investigation, which started in 1951 and is now coming to an end. We then briefly summarize the main properties of consanguinity, inbreeding, and drift, as well as inbreeding effects in humans and some special projects that were part of the investigation. Finally, we summarize the structure of this book.


## 1.2 HISTORY OF THIS RESEARCH


In 1951 Luca Cavalli-Sforza started teaching a course in genetics at the Faculty of Sciences of the University of Parma, Italy. Among his students was Antonio Moroni, a priest who is now professor of ecology at the University of Parma. At that time Moroni taught natural sciences at the Seminary of the Parma Archbishopric and made Cavalli-Sforza aware of records in the Roman Catholic archives that could be of interest for human genetics: essentially dispensations for consanguineous marriages and parish books of deaths, marriages, and baptisms. In an almost 100% Catholic country, baptisms are a close equivalent of birth records. Newborns are usually baptized very soon after birth, and a very small fraction, probably less than 1%, dies without a chance to be baptized. Moroni was also instrumental in obtaining permission from the higher religious authorities to use these records for genetic investigations. Our investigations began at the bishopric of Parma. They were soon extended to other bishoprics, and eventually to the whole country of Italy. To help with our investigations, a letter from the highest Catholic authorities was sent by the Vatican to all parish priests,* asking them to make parish records

---

* A letter dated 15 December 1960 by His excellency Monsignor Cesare Zerba,

available for scientific purposes. Genetic research using Roman Catholic records was also started in France by Jean Sutter and his collaborators, at about the same time as ours.

Consanguinity records are to be found in various Catholic archives. Consanguinity itself is very carefully defined and Roman Catholic legislation prescribes with great precision which marriages are completely forbidden, which ones are permitted under dispensation from a higher religious authority, and which do not require dispensation. Priests receive formal teaching about these rules in seminaries in which they also learn to evaluate accurately the degree of consanguinity of candidates for marriage. The need for dispensation has changed through time and is now reduced to a minimum. In earlier times even remote degrees of consanguinity were forbidden and it was essential to ask for dispensations before marriage could be celebrated. A consanguineous marriage celebrated without the requested dispensation would be null and void, generating a very serious social problem for the families.

Chapter 2, on the history of consanguinity regulations in the Roman Catholic Church, examines the historical knowledge available. There are also geographic differences in rules for obtaining dispensation. In peninsular Italy the parish priest must check every pair of prospective spouses for the existence of recent relationship, and if one requiring dispensation is discovered, he must request it from the bishopric. A copy of the request is then sent from the bishopric to the Vatican, and is returned to the bishopric with the Vatican response

---

Secretary of the Congregation for Sacraments, gave permission for the use of consanguinity dispensations held in the Archives of the Congregation of Sacraments and suggested that Italian bishops would allow the use of consanguinity dispensations held by Diocesan Archives, if the need to complete the investigation would arise:

The Reverend Dr. Antonio Moroni, Professor at the University of Parma, intends to investigate the consequences of marriages among consanguineous persons and among minors that celebrated their wedding after receiving dispensation from their otherwise forbidding condition. The Congregation of Sacraments, believing that the results of the investigation will help better understand the Church norms that discourage the celebration of such weddings, grants permission to Reverend Prof. Moroni to have access to the needed data that are held in the Archive. For similar permission to access the Diocesan Archives he will also address their excellencies the Italian Bishops, who most certainly will not miss the social, pastoral, and juridical usefulness of the scientific investigation started by the University of Parma.

and then to the parish before a marriage requiring dispensation can be celebrated. In a few regions other than peninsular Italy, dispensations for at least some less close consanguinities could be given by a local Catholic authority other than the Vatican.

At least in peninsular Italy, therefore, there are three sources of records: the priest is supposed to indicate on the parish marriage book that a specific marriage required a dispensation, but we found that in some parishes this was not always carefully done. Folders keeping full records of dispensations are kept in each bishopric, and a slightly less complete set of records is available in the Vatican archives. It is rare that a dispensation is not approved by the Vatican, if it is dispensable. Genealogical trees of the candidate spouses reconstructed by the parish priest are extremely common in the bishopric archives that we investigated, and are available for practically every dispensation requested, but much less frequently in the Vatican archives. These genealogical trees are essential for checking the consanguinity degree calculated by the priest and for testing hypotheses on age and migration effects, to be described in a later chapter. The presence of genealogies in the dispensation folders made it possible to check for errors in the calculation of consanguinity degrees. Errors were nonexistent or exceptional, not surprisingly, since the method of computation is regularly taught to priests at seminaries.

Our work was based initially on bishopric records. After a full study of dispensations deposited in the archives of the Archbishopric of Parma—the diocese of the city where the university in which Cavalli-Sforza taught from 1951 to 1962 is located—those of the two adjacent dioceses of Piacenza and Reggio Emilia were also investigated. The territory of these three dioceses and cities has almost identical ecological structure, extending from the Appennine mountains to the lowest part of the plain of the Po River. The three dioceses form the northern moiety of the administrative Italian region called Emilia, practically at the center of the Po valley, in the northern part of Italy. The Po River flows just north of the city of Piacenza, and continues eastward toward the Adriatic Sea. Parma and Reggio Emilia are on a major Roman road, the *Via Aemilia*, an almost straight line in the Po valley leading east-southeast from Milan to Bologna. The region around the Po and the *Via Aemilia* is a very fertile plain. Proceeding from each of the three cities toward the south, one enters first a hilly region and then a mountainous one. Population density is maximum near the cities, which are all located not far from the center of a very prosperous agricultural region. The mountainous region at the southern end of each diocese has the lowest population density. It is a part of the Appennine chain, the crest of which separates Emilia from Tuscany. The hilly region, intermediate between the plain and the mountains, has intermediate population density. The size of villages is on average highest near the cities and decreases regularly toward the mountains. This variety of environments of each diocese has helped in the investigation of the effects of the relevant ecological and demographic variables.

Bishopric records of the islands of Sardinia and Sicily, as well as of some other islands and inland regions of special interest, were also investigated, showing similar effects of demographic variables, along with other characteristics of each region. Records of individual dioceses of the islands were published earlier; their analysis has been repeated by partially new approaches for the purpose of preparing this book.

In later years it was possible to establish a team of young students who copied the consanguinity records of the Vatican archives under direction of Father M. Bracco. These records included all of peninsular Italy from 1911 to 1964, except for Sicily, which had independent rules. It was necessary to visit directly the bishoprics of Sardinia and Sicily, but our survey of Sicilian dioceses was not complete.

Full names of consanguineous couples were available. They were kept confidential, but we had permission to use them for linking them with other records to study the effect of consanguinity on certain phenotypes. The records were eventually transferred to computer tapes, analyzed statistically, and ordered alphabetically.

Results on some of the bishoprics were in part published earlier. But the major analysis, that of the Vatican records, is published for the first time in this volume. A number of other new calculations were carried out on the available records and are included in this book. Other socioeconomic investigations done in Italy were studied and correlated with the consanguinity data. Socioeconomic information came from the Istituto Centrale di Statistica.

Closely connected with the analysis of consanguinity, studies of demography from parish books, mostly of the Archbishopric of Parma, were started. We realized that the parish books gave us access to demographic evaluations that could be used for comparing the amount of observed and expected "random genetic drift" (genetic variation between villages generated by chance effects, calculable on the basis of information on village size and migration among villages). At the time, the importance of drift in determining genetic variation in humans was considered by many geneticists to be trivial without real proof. Here was thus an opportunity to compare expected and observed genetic variation in real cases, and possibly to evaluate the relative role of drift versus other evolutionary factors. Genetic data were obtained on blood groups of inhabitants of most of the 92 villages of the valley of the Parma River, and correlated with the genealogical studies that form the basis of the present book. Many other investigations have since been done following the same scheme, but none is as large and complete. Only summaries of the Parma valley drift investigation have been published until now; full data are available for the first time in this volume.

Further use of the data collected in the consanguinity studies was made possible by methods we developed for studying surnames as genetic markers. Very recently it has been established that the Y chromosome, which is transmitted from fathers to sons and the presence of which determines the male sex, provides an extremely useful set of genetic markers. The differences in migration at marriage between males and females, mentioned earlier, make genetic diversity of the Y chromosome greater than that of the other chromosomes. Until recently, very few genetic markers of the Y chromosome existed, but this situation is now changed (Underhill et al. 1997, and later papers). Surnames, however, are Y-chromosome markers—they have on average a much younger age than DNA mutants, but are very useful for certain purposes. Parish books and consanguinity data have provided us with a number of surname data that could also be used for the study of genetic variation and give valuable estimates of drift. Thus, we could examine in detail the strong correlations existing between three major phenomena: consanguinity, inbreeding, and drift. We will examine these correlations in some detail.

## 1.3 CONSANGUINITY

The word consanguinity derives from the mistaken notion that blood (in Latin "sanguis") is the basis of inheritance. Two consanguineous people have at least one ancestor in common. In principle, one can prove that any pair of human individuals has common ancestors, but to find those of two individuals taken at random one would usually have to go back for a great number of generations. It is thus necessary to introduce a limit to how ancient common ancestors can be for considering two individuals as consanguineous. This limit varies with custom and is, of course, arbitrary. There is, therefore, no fixed limit. When Cavalli-Sforza and Bodmer discussed the problem in chapter 7 of the book *The Genetics of Human Populations* (1971, 1999), they mentioned as an example that the bond of two individuals who have a great great-great-grandparent in common (and are, therefore, as we shall see, fourth half-cousins) is very tenuous, indeed. The probability that they share a gene they inherited from their common ancestor is about one in a thousand. The most remote consanguineous pair usually considered is that of third cousins, full or half, for which the probability of sharing a gene transmitted from common ancestors is 1/256 or 1/512, respectively (see figure 1.1).

The case of half-cousins (descending from only one common ancestor) is less frequent than the normal case of full cousins. The common ancestor of half-cousins usually married twice and the two consanguineous mates descend each from one of his or her two spouses. Consanguinity can be simple or multiple, depending on whether there is only one chain of descent from a couple of married common ancestors or there are chains of descent from more than one couple. For instance, two consanguineous spouses can be first cousins from one pair of ancestors and second cousins from another.

The most important simple consanguinities are shown in figure 1.1. The three relationships involving equal length of the two branches leading to the husband and wife (1st, 2nd, and 3rd cousins) are usually the most common degrees. Roman Catholic terminology is shown in the first column of degrees of relationship. Civil law definitions usually follow the Napoleonic code, given in the next col-

| Type | Symbol | Degree of Relationship | | Inbreeding Coefficients (F) | |
|---|---|---|---|---|---|
| | | Roman Catholic Usage | Napoleonic Code | Full | Half |
| Uncle-niece; aunt-nephew | 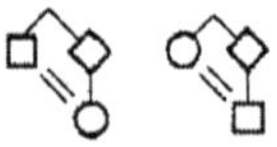 | I in II | III | 1/8 | 1/16 |
| First cousins | 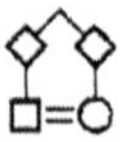 | II | IV | 1/16 | 1/32 |
| First cousins once removed($1\frac{1}{2}$) | 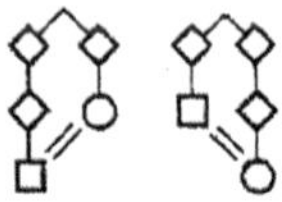 | II in III | V | 1/32 | 1/64 |
| Second cousins | 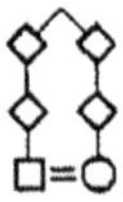 | III | VI | 1/64 | 1/128 |
| Second cousins once removed ($2\frac{1}{2}$) | 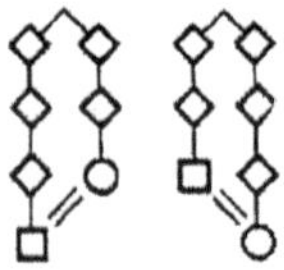 | III in IV | VII | 1/128 | 1/256 |
| Third cousins | 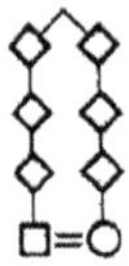 | IV | VIII | 1/256 | 1/512 |

FIGURE 1.1. The most common types of consanguineous matings, their symbols, and inbreeding coefficients: □, male; ○, female; ◇, an individual of either sex. "Full" and "half" refer to the two sibs starting the chains of descent, who are the two top individuals in each pedigree. Full sibs have both parents in common, and half-sibs only one parent. *From Cavalli-Sforza and Bodmer [1971, 1999].*

umn. The measurement of genetic similarity among spouses is expressed by the inbreeding coefficient $F$, shown in the last two columns for half- and full cousins. Its meaning is described in the next section.

## 1.4 INBREEDING MEASUREMENT

An individual who received from his parents two identical forms of a given gene is said to be homozygous (or a homozygote). In a heterozygote the two gene forms of paternal and maternal origin are different. The indication that an individual is homozygous or heterozygous refers specifically to a single gene (traditionally, a DNA segment with a specific function, but sometimes a DNA segment specified in another way). Different forms of a gene are called *alleles*. Until one could examine DNA directly the number of detectable alleles was very small. Our current ability to examine DNA directly allows us to distinguish all mutations that have occurred and are maintained in the population examined. Today the physical identity of two genes can be ascertained with practical certainty by appropriate tests of DNA. This is called *identity by state* or *identity by nature*.

Inbreeding may cause another type of identity, *identity by descent*, when two descendants receive the same gene from a common ancestor, but his/her two alleles of that gene are not necessarily identical by state or nature. The possibility that two genes identical by descent are not identical by state arises if a gene passed from a common ancestor to two descendants undergoes mutation in at least one of the two chains of descent, leading to the two descendants. When this happens the particular gene being considered will be different by state in the two individuals, although it is identical by descent. This distinction is important for theoretical work.

Some semantic confusion may be generated by the fact that the word "gene" is used to indicate both loci and alleles. Historically, loci (plural of locus) are positions in a chromosome, as established by studies of crossing over. The word "locus" is used more loosely today. In genes of known sequence, a locus may be the position of a single nucleotide, the smallest element of DNA (sometimes a longer segment), but this is more properly called *nucleotide site*. Alleles are the alternative forms of DNA observed at a given locus. They often differ one from the other because of a single nucleotide, but sometimes because of a longer segment. There is possible confusion between alleles defined at the level of a whole gene (a DNA segment with a precise function) and at the level of DNA. In pre-DNA times the distinction among alleles was not as sharp as is possible today, when we know the DNA structure of a gene. All differences in state giving rise to different alleles are caused by one or more past mutations.

An individual homozygous by nature need not be homozygous by descent, and usually is not. The test of inbreeding by nature may require knowledge of the genealogy for very many generations to establish common descent. In principle, one can also distinguish homozygosity by descent from homozygosity by state also empirically, by looking at nearby genes. There usually is variation in DNA very close to that being studied; genes identical by descent, even if they are not identical by nature, will have identical neighbors, except for earlier mutations, which are not frequent; genes identical by nature but not by descent may have different neighbors. This diagnosis requires thorough advanced knowledge of neighboring genes at the DNA level, ideally the full sequence.

The inbreeding coefficient $F$ measures the probability that an individual receives at a given locus two genes identical by descent, calculated on the basis of the relationship of his/her parents. It is the same for all genes of autosomal chromosomes and can be calculated from the pedigree of an individual. Figure 1.2, taken from figure 7.1 of Cavalli-Sforza and Bodmer (1971, 1999), shows the calculation for a simple example, a half-uncle–niece marriage, which has $F = 1/16$. The genes of the uncle (A) are labeled $a$ and $b$, while those of his two spouses are assumed to be different from $a$ and $b$ and between themselves. As there is no need to distinguish them, they are all labeled $+$. The two children B and C can be either $a/+$ or $b/+$, with probability 1/2. If C is $a/+$, then D, the niece of B marrying him, is also $a/+$ with probability 1/2. If D is $a/+$ and B is $a/+$, then their child E can be homozygous ($a/a$) with probability of 1/4. Altogether, the probability that E is $a/a$ is the product of 1/2 (the probability that B is $a/+$) times 1/2 (the probability that C is $a/+$), times 1/2 (the probability that D is $a/+$ if C is $a/+$), times 1/4 (the proba-

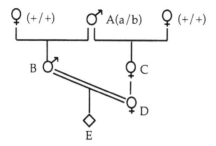

FIGURE 1.2. Half-uncle–niece marriage. Coefficient of inbreeding $F$ = 1/16.

bility that a child of B and D is $a/a$ if both are $a/+$ ). This product is 1/32. There is an equal, independent probability of having a $b/b$ child. The total probability that the child is homozygous $a/a$ or $b/b$ is the sum of the two probabilities, $1/32 + 1/32 = 1/16$.

Full uncle–niece (UN) marriages have twice that value, and aunt–nephew (AN) marriages give the same value, 1/8 (these values differ for X chromosomes). Adding a generation to one of the two chains of descent leading to the spouses multiplies $F$ by 1/2. For more complex pedigrees a simple method of computation is shown in most population genetics textbooks (see also Cavalli-Sforza and Bodmer [1971, 1999], chapter 7, from which this example and many of the following ones in this chapter are taken).

## 1.5 INBREEDING EFFECTS

Inbreeding increases homozygosity above the values expected under random mating in a population of finite size. Frequencies of genotypes of ordinary populations are predictable under assumptions of random mating in an infinite population. These give rise to the *Hardy-Weinberg* distribution of genotype frequencies. For alleles $A_1$, $A_2$, . . . , $A_k$ with frequencies $p_1, p_2, . . . , p_k$ this distribution is given by

$$(p_1A_1 + p_2A_2 + \cdots + p_k A_k)^2$$

The result is that homozygotes $A_1A_1$, $A_2A_2$, . . . , $A_kA_k$ have frequencies $p_1{}^2, p_2{}^2, . . . , p_k{}^2$, and heterozygotes $A_1A_2$, . . . , $A_jA_k$, . . . have frequencies $2p_1p_2, . . . , 2p_jp_k, . . . .$

But individuals with inbreeding coefficient $F$ have expected frequencies $Fp_i + (1 - F)p_i^2$ if homozygotes, and $(1 - F)2p_jp_k$ if heterozygotes $A_jA_k$. Therefore, all heterozygotes have a lower expectation under inbreeding than under random mating by a factor $1 - F$, and homozygotes a correspondingly higher one. If a population is completely inbred, then $F = 1$, and only homozygotes exist.

The average inbreeding coefficient of a population is called $\alpha$ and an approximate estimate can be obtained by averaging the inbreeding coefficients of its individuals. If the frequencies of consanguineous matings in the population $m_i$ and their inbreeding coefficients $F_i$ are known, then

$$\alpha = \Sigma m_i F_i$$

$\alpha$ values vary considerably with space and time (see www.igm.cnr.it/Zei/freqcons.html).

One usually neglects in these calculations more remote consanguinities, because they are not known, and it is inevitable to wonder if this computation leads to serious underestimates. An exact answer is difficult. It might seem that in most instances the inaccuracy is not serious, given that third cousins already contribute almost negligibly to the final estimate. The inaccuracy is likely to be more serious in highly inbred populations that are very small and have remained small for a very long time, so that inbreeding is more likely to have accumulated over time. One of the highest $\alpha$ values (0.044) was found in Samaritans, who number only a few hundred individuals and have essentially no gene flow from outside populations. Small religious isolates in North America, who also have negligible gene flow from outside, except for other colonies of the same denomination, show $\alpha$ values up to 0.025. In one case (Mange 1964) some idea of the underestimation thus incurred in a growing human isolate was possible, and was not entirely trivial. The fact that this isolate was growing rapidly introduced a further factor, however. We have already mentioned that careful avoidance of inbreeding in a very small group of Polar Eskimos has kept $\alpha$ rather low (less than 0.003 [Sutter and Tabah 1956]).

Deviations from Hardy-Weinberg (HW) in human populations are rare. The extent of deviations caused by inbreeding is rather small, making it unlikely that an inbreeding effect will manifest itself by deviation from HW. Natural selection is more likely to cause the opposite deviation, making heterozygotes more frequent than expected. But admixture of populations with different gene frequencies, as is unavoidable in very large samples collected over wide geographic areas, can cause a relative decrease of heterozygotes, simulating nonexistent inbreeding. Some improperly call inbreeding the loss of heterozygosity due to recent admixture, as can be observed in heterogeneous populations. In general, deviation from Hardy-Weinberg is not a useful cue for detecting inbreeding effects, as it is more likely that it is the result of admixture of populations because of poor sampling schemes.

Especially in plants, but also in animals, extreme inbreeding may cause loss of fertility, resistance to diseases, and phenotypic deterioration (e.g., loss of height). This happens, however, on $\alpha$ values higher than those observed in humans. Thus, inbreeding depression, as the phenomenon is called, is only moderate in humans. The very low fertility observed in various indigenous groups of the Andaman Islands, who are now reduced to less than 100 individuals, may be one of the worst cases of inbreeding depression, but the $F$ is not precisely known in the absence of long genealogies, and there are other possible causes of loss of fertility.

Ways of analyzing inbreeding depression in humans include the study of *genetic load*. Genetic load is defined in terms of Darwinian fitness, which is the number of children contributed to the next generation by a genotype. Darwinian fitness is usually but not always expressed as relative to that of the genotype with highest fitness (Crow 1958). The genetic load, as first defined by H. J. Muller (1950) is the loss of Darwinian fitness due to deleterious genes maintained in a population by the balance of adverse mutation, producing deleterious genes, and natural selection eliminating them.

New mutants are sometimes deleterious. The relative loss of Darwinian fitness of the homozygotes for a specific mutation, compared with that of the normal homozygotes, is called $s$, and that of a heterozygote for the normal and mutated form of the gene is $hs$. To avoid possible confusion, this means that we call $1$ the fitness of the homozygous normal, $1-hs$ that of the heterozygote, and $1-s$ that of the homozygous mutant. Therefore, $h$ is a quantity specifying the dominance of the mutation, being 0 if the mutant is fully recessive, 1 if it is fully dominant, intermediate between 0 and 1 if dominance is intermediate, and negative if there is heterozygous advantage (also called "overdominance"). Mutation at that locus, occurring at rate $\mu$ per generation, keeps accumulating new mutants, which are eliminated at a rate increasing with $s$. At equilibrium between mutation and selection the number of new mutants per generation is equal to those eliminated by natural selection. The proportion of mutant genes at that locus existing in the population at equilibrium depends on mutation rate $\mu$ and the selection parameters $s$ and $h$. If $h = 0$, the mutant is fully recessive and the gene frequency of the mutant in the population at equilibrium is $\sqrt{\mu/s}$. The mutant phenotype before selection has frequency $\mu/s$, that is, it is equal to the mutation rate if the mutation is completely lethal ($s = 1$); otherwise, it is higher. If the mutation is completely dominant ($h = 1$), then its equilibrium gene frequency is $\mu/s$ and the frequency of the phenotype is twice that value. When the heterozygote is intermediate between 0 and 1, its fitness tends to dominate the picture even if $h$ is small, because heterozygotes for rare mutants have a frequency much higher than that of homozygotes, which are usually very rare.

Morton et al. (1956) applied the Muller approach to estimate the genetic load in human populations. The load $L$ is the loss of Darwinian fitness and is a linear function of the inbreeding coefficient, being, for a specific locus $i$,

$$L_i = a_i + b_i F$$

where $L$ can be calculated as minus the logarithm of mortality (or other measurements of load).

Summing over all $i$ loci gives

$$L = A + BF$$

provided different loci act independently, so that the joint probability of different damages is their product.

The expected linear relationship was tested in a large number of populations. The best data set seems to be one of the oldest, produced by Schull and Neel (1965) from estimates obtained in Nagasaki and Hiroshima. The mortality examined was that of children from first-, 1½-, and second-cousin marriages, the most common and most easily detectable consanguineous marriages, which are particularly frequent in older Japan. Here it was found that $B = 0.4$ and $A = 0.04$. Note that the mutation estimates on which these data are based are not related to radiation effects, but express mutants produced in earlier generations, long before atom bombs were used against these two cities during World War II.

The quantity $B$, or a slightly larger one, included between $B$ and $A + B$ is approximately equal to the sum of $sq$ over all independent loci contributing to mortality, where $q$ is the gene frequency of a deleterious allele at a given locus, and $s$ is the loss of fitness of the mutant homozygote. It has been called "the number of lethal equivalents," that is, the number of lethals that would be found in gametes, assuming that $s = 1$ for all genes. Since $s$, the mortality due to a deleterious but not fully lethal gene, is smaller than 1, the number of deleterious genes in the gametes of an average individual from the population must be greater than the number of lethal equivalents.

The comparison of different populations shows considerable variation of $A$'s and $B$'s. An old summary (Cavalli-Sforza and Bodmer 1971, 1999) indicated that Japanese values are lower than average; data from Caucasoids tend to be more variable and higher, varying from less than 0.5 to more than 2.5. Limiting the analysis to first-cousin data, which are the most reliable, Bittles and Neel (1994) estimated the mortality due to recessive deleterious genes to be around 4.4%, from which one can calculate that the average person carries 1.4 lethal equivalents.

A recent, thorough meta-analysis by Grant and Bittles (1997) proved that there is considerable heterogeneity among the 42 populations considered. Linearity with $F$ was not tested directly by the authors, but the data they published indicate a serious deviation from linearity at the upper end: the highest degrees tested (uncle–niece or equivalent, i.e., double first cousins, which have $F = 1/8$) had a mortality lower than twice that of first cousins (twice is its approximate expectation). Even if mortality was not perfectly linear with $F$, there was a clear effect of consanguinity, which was qualitatively in the expected direction. There were no obvious differences in the sensitivity to $F$ among the different types of mortality: prenatal and postnatal in various age ranges, including postreproductive ones.

The interpretation of load analysis is not as straightforward as was suggested in the first studies. Theoretically, it can be shown that it is difficult to distinguish *mutational load*, as the H. J. Muller type of load is called, from *segregational load* (due to polymorphisms responsible for heterozygous advantage). Also, possible nonadditive interactions between genes make the estimate of lethal equivalents uncertain. Nevertheless, if one can confirm that the load estimated in Japanese populations is truly lower than in others, which have had less consanguineous marriages, and if one can assume that the Japanese custom of frequent consanguineous marriage has been going on for a substantial time, it is reasonable to conclude that this practice has freed this population from part of its mutational load. A lethal recessive mutation that has taken place in the past will sooner or later determine one genetic death in a stationary population (and more in a growing population). The regime of consanguineous marriage will affect the delay with which this happens.

## 1.6 RANDOM GENETIC DRIFT

Drift is the name given by Sewall Wright to the effect of chance on gene frequencies in successive generations. Higher organisms form the next generation by sexual reproduction: gametes unite to form zygotes by fusion of a spermatozoon and an egg cell. If $N$ individuals are formed from those of the earlier generation, $2N$ gametes of that generation must have been used to form the new generation, and they may be viewed as a random sample (in the absence of gametic selection) of the genes present in the earlier generation. Under these conditions, the genes of every new generation are a random sample of those of the last generation, and one can compare the formation of the next generation to the taking of a random sample of gametes from the former generation. The nearest statistical analog of this sampling is called "binomial sampling," which is like drawing a sample of black and white balls from a bag where they are contained in known proportions. The two quantities of importance are the proportion $p$ of black or white balls in the bag and the number $N$ of balls taken out to form every new generation. Actually, there are two bags: one of male gametes and one of female gametes, but in practice they almost always have the same composition. To simulate the very large number of gametes from which are taken the few sorted to form the next generation, one always puts the black or white ball back into the bag after its extraction, before taking another ball. This is the same as "binomial sampling"; it is not realistic but the deviation thus generated is negligible.

The important conclusion to take home is that chance affects long-term results of evolution, the more so, the less numerous the individuals of a population. A population bottleneck can have drastic effects, especially on pathology. If a population is generated by very few founders, say 10, and one of them happens to have a rare inherited disease, the frequency of that disease will be 10% after the bottleneck and will tend to remain 10% if the diseased individuals are not subject to strong adverse selection. This is especially true of diseases that do not cause prereproductive mortality or loss of fertility. The frequency of the diseases will oscillate, of course, statistically, in successive generations. There are many examples of populations that have abnormally high or low frequencies of certain genetic diseases, specifically because of drift effects due to a history of one or more demographic bottlenecks in the recent past, or to a persistently low population size. The variation may be in both directions: the disappearance of a disease or the presence of it at an abnormally high frequency. Every population will therefore show a different picture of incidence of genetic diseases, which will deviate more from average, the smaller the bottleneck at founding of the population or at any later time. The effect will vary enormously in the same population for different diseases.

Some people have confused drift and inbreeding. These are really two different phenomena, but there is a strong correlation between the two. When the population size $N$ is small, the chance is that one will mate more easily with a consanguineous individual. Small $N$ is likely to generate inbreeding, and of necessity it also generates drift. Thus, a population of a given small size $N$ whose individuals marry at random may be affected by a certain degree of inbreeding even if mating is random (independent of degree of consanguinity of mates). We may call this a *random inbreeding*. But mating systems may keep the inbreeding level high or low with respect to that expected under random mating: if consanguineous matings are avoided, then inbreeding is reduced; and if they are favored, then inbreeding will increase. One of the aims of the present research is to find whether systems of mating of human populations cause the degree of inbreeding to be higher or lower than expected on the basis of population size.

Use of the word drift in genetics is equivalent to the long-term effect of statistical fluctuations of gene frequencies due to the finite size of populations. The concept involves, therefore, only the effect of random events. But the word drift is used with an opposite meaning in other sciences, like physics and linguistics, where it usually means a definite trend in specific directions. The *Oxford English Dictionary* suggests that drift means a definite tendency: the act of driving in a specific direction, or the condition of being driven, as under the action of a current. Kimura has therefore suggested the use of a longer expression, "random genetic drift", when it is useful to avoid potential confusion.


## 1.7 RESEARCH ON DRIFT IN THE PARMA VALLEY

The Parma River starts in the Apennines and descends northward toward the Po, the major river in Italy, into which it flows at the center of the northern plain. The valley it forms is separated to some degree from the two parallel valleys, east and west. Its geographic structure seemed ideal for a study of the effect of random genetic drift. It has a multitude of villages that increase regularly in size, on average, as one descends from the highest altitude toward the lowest. Village sizes, and genetic exchange by migration between them, mostly due to marriage, have been relatively stable over the centuries. Almost every village had its parish church, and a great majority of the parishes were already in existence in the eleventh century. Only a few have been founded since that time. Since the end of the sixteenth century, parish books of baptisms, deaths, and marriages were kept in each parish church and it became clear that they could be made available for study of the demography of the region.

In 1955, Cavalli-Sforza decided to begin an analysis of drift in the Parma valley. The plan was to study the demography of the valley on parish books and evaluate theoretically on the basis of these data the amount of drift expected in the various sections of the valley. At the same time, blood samples would be obtained from at least 30–50 adults, as unrelated as possible, from the parishes, and the only genetic markers available at the time, major blood groups, would be tested on them.

Data collection lasted until the end of the 1950s. Statistical analysis went through various phases. It was soon clear that, by and large, the genetic variation of the Parma valley could be explained on the basis of drift, but there were problems tied with the existing methods of analysis. In the 1960s many methods were tried and new methods proposed, including what was probably the first use of population simulations. These difficulties were not completely resolved at the time the book *The Genetics of Human Populations* by Cavalli-Sforza and Bodmer was published (1971, reprinted without change by Dover in 1999). Eventually it was concluded that the most satisfactory method of analysis was population simulation, and final proof that drift could fully explain the results is published in this book.

## 1.8 GENETIC USES OF SURNAMES

It has long been clear that surnames, being transmitted from fathers to children, contain information potentially of interest for genetics. In particular, in consanguineous marriages both spouses have a specific probability of having the same surname (isonymy). The study of isonymy of spouses has been considered as a method of estimating inbreeding and kinship in a population (Crow and Mange 1965). The relationship is complicated by the fact that some surnames are very frequent, probably because the same surname arose many times independently (polyphyletism); therefore, the identity of surnames of two individuals is no proof that they are related. In general, the method needs corrections on the basis of the frequency distribution of surnames in the specific population, and has shown variable appreciation, depending on authors and on specific populations investigated. Lasker's book (1985) is dedicated especially to the use of surnames, with special but not exclusive interest in isonymy, and its usefulness for investigating genetic population structure. Two other books (the first arranged by K. Gottlieb [1983], the second by Brunet et al. [2001]) are collections of papers using the method and evaluating it, as well as other ways to use surnames for the same purpose.

We chose to avoid the use of isonymy because our interest in

surnames was aroused by other approaches. The rationale is that surnames can be considered as a single gene with many alleles. Such
genes are particularly informative for evolutionary purposes, and may
be very useful because many surname data on human populations are
already available and computerized. They differ from the usual genes
(with the exception of those of the nonrecombining portion of the Y
chromosome, which behaves in a similar way) in being almost universally transmitted by only one parent, of the male sex only. They
therefore are transmitted as in unisexual haploids, a fact requiring
some simple corrections in the formulas to be used for estimating
evolutionary factors like migration and drift. One weakness is that
surnames are affected by illegitimacy to an unknown extent; this will
generate a deviation from the usual gene behavior, which can be
formally expressed by considering them as Y-chromosome "genes"
affected by a high mutation rate. Another limitation of surnames is
that they arose mostly recently, and therefore genealogies based on
them have validity only for a short lapse of the past. This reason is to
some extent confounded with problems due to illegitimacy. However,
even in countries where they arose as recently as in Japan during the
Meiji restoration (1867–1912) (Yasuda and Furusho 1971) they have
been fairly useful for some genetic purposes.

   We have used surnames especially to evaluate two types of genetic analysis of populations: random genetic drift and migration.
The quantities involved are population size $N$, and the migration rate
$m$ per generation (the proportion of individuals of the population that
entered from other populations in the last generation). Both quantities
are necessary for evaluating the effect of "population structure" on
evolution. The effect of drift is stronger the smaller $N$ and the smaller
$m$. In fact, their product, $Nm$, can be used as an inverse measure of
drift, which increases as $Nm$ grows smaller. Naturally, these quantities are bound to change over time, and what would be necessary is
their knowledge over an adequate time period, but this is rarely available. In this respect, records from the Catholic Church are valuable,
but their examination is time-consuming.

   One of the constant and difficult problems in the study of population genetics is the separation and evaluation of the relative roles of
genetic drift and natural selection as they affect the evolutionary
change of gene frequencies over time and space. When genes are

studied it is not always easy to distinguish the effects of these two major evolutionary forces. But one can expect surnames to be relatively unaffected, per se, by natural selection, and this is of some advantage with respect to other traits.

The estimation of population size $N$ is easier for human populations than for most other living organisms, and for this reason human populations are among the best organisms for the study of drift. To some extent, the study of migration is also easier in human populations, although it is less easy to estimate than population size. For certain purposes one needs a special formula for the calculation of $N$, known as effective population size $N_e$, but in many cases it is not necessary. When this estimation is necessary, in human populations the effective population size is approximately one-third of the global population, as explained in Cavalli-Sforza and Bodmer (1971, 1999). In the case of surnames, $N$ should ordinarily be limited to the number of males. When this is strictly necessary we will make sure that the refinements of using $N_e$ and $N$ of males only are followed, but when examining correlations or studying proportionality to other demographic quantities neither restriction is important. One can obtain a quantity very similar to $Nm$ from a formula involving the number of different surnames and the number of individuals in a population unit. This formula was developed by Fisher (1943) for other purposes, but we have started applying it in population genetics.

The second use is for calculating genetic variation among populations, with formulas taken from the $F_{ST}$ variance of gene frequencies. This is another approach to the study of drift and, more generally, genetic variation among populations. Variances are calculated independently for each surname and averaged over surnames.

The first collection of surnames we used was from consanguineous marriages. We had computerized, for other reasons, around 540,000 consanguineous marriages for which there existed, in the archives of the Vatican, a consanguinity dispensation for the years 1911–1964. The archives excluded Sardinia and Sicily, which were done independently. Later, we were able to obtain surnames on magnetic tapes from telephone books by the SEAT Company for all Italy and, for Sardinia, also from electricity bills of households by the ENEL Company. Telephone listings for the year 1978 were examined for 91 Italian provinces and first used for a study of the migration per province. We compared the migration that could be obtained from the quantity $Nm$ derived from each province from surnames and the migration measured directly by the Istituto Centrale di Statistica (ISTAT). There was good agreement (Piazza et al. 1987). In several other papers (Zei et al. 1983a,b, 1986, 1993, Wijsman et al. 1984, Lisa et al. 1996) we studied specific problems like estimation of genetic parameters, relations with drift and natural selection, and, in general, the use of surnames for genetic problems.

We later extended the analysis of surnames for genetic drift to the approximately 8,000 Italian communes, again using telephone books by SEAT Company from 1993. Conclusions and maps are given in chapter 10. We also carried out several other correlations to cross validate data obtained on surnames and on regular genes, whenever a direct comparison was possible. This analysis confirmed the usefulness of surnames for most genetic analyses. Here we consider especially their use for the analysis of drift and for studying correlations of drift with inbreeding and consanguinity.

## 1.9 A SUMMARY OF PUBLISHED STUDIES ON CONSANGUINITY AND INBREEDING, WITH SPECIAL REFERENCE TO ITALY

The first published data on consanguinity and kinship in Italy are from Mantegazza in 1868. His pioneer work entitled "Studj sui matrimonj consanguinej" was based on the analysis of the offspring of 512 families (90 in Italy) with different degrees of consanguinity. It was only in the 1930s that there appeared other studies on consanguinity in Italy. As reported by Serra and Soini (1959), these studies regarded isolated areas, such as small communities, but also entire alpine valleys (Cantoni 1931, 1935, 1936, 1938, Gianferrari 1932, 1936). Since World War II, this research has spread.

Sometimes the study of the effects of consanguinity is the reason for collecting great samples of data on marriages often over long periods. Fenoglio (1956, 1969) searched for a relationship between consanguinity and sterility and between consanguinity and blood groups, analyzing all the marriages that occurred in a province of North Italy (Cuneo) during 1901–1960. Bigozzi et al. (1970) studied morbidity and mortality in the offspring of 300 consanguineous marriages celebrated in Firenze during 1939–1958, compared with an equal number of nonconsanguineous marriages. Recently, Danubio et al. (1999) searched for a relationship between inbreeding and malaria in the southern region of Calabria. Marriage behavior, in particular, the rate of endogamy and inbreeding, is being studied by Guerresi et al. (2001), who are analyzing 4518 marriages that occurred in 7 parishes of the alpine Non valley.

More often, the studies were essentially devoted to analyzing frequency and trend over time of consanguineous marriages in some particular area or in the whole of Italy and searching for demographic and socioecomic factors affecting them. Serra and Soini (1959, 1961) examined all the marriages celebrated in three provinces of North Italy (Milano, Como, and Varese) during the period 1903–1953. Barrai and Moroni (1965) gave a preliminary picture of the consanguinity trend in the Reggio Emilia diocese during three centuries (1631–1963) through the analysis of a sample of parishes corresponding to 23% of the total. A long period (1565–1980) was examined for consanguinity in the Upper Bologna Apennine by Pettener (1985), who found the frequencies and trend of consanguineous marriages to be very close to those of the other Emilian dioceses of Parma, Piacenza, and Reggio Emilia, described in this book.

All these studies were based on data collected in ecclesiastic archives in the parishes or dioceses. These data are believed to be reasonably accurate for reasons described in this book (chapter 3). Not so reliable are the other sources of consanguinity data, as, for example, official statistics. Fraccaro (1957) made one the first studies of consanguinity in all regions of Italy based on the data provided by the Istituto Centrale di Statistica for the year 1953. The values of the regional inbreeding coefficient are underestimated by a factor of about three, compared with those obtained from the ecclesiastic sources. The author himself looks with caution at the results, remembering that "the data furnished by the official statistics on consanguinity depend, on one hand, on the *bona fides* of the partners and, on the other, on the *bona voluntas* of the officials" (Serra 1959).

Somewhat more reliable seems the estimate of consanguinity frequencies obtained by Cavalli-Sforza (1960) through the "Special investigation on the consanguinity of the marriages" that was coupled to a survey of the Istituto Centrale di Statistica in 1959. This pilot study performed on a sample of 5‰ of the Italian families gave values of consanguinity underestimated in comparison with those of ecclesiastic sources, but only by a factor of about 1.5.

A bibliography of recent investigations of consanguinity frequency, and of its effects, outside Italy is available from Alan Bittles at www.consang.net.

## 1.10 STRUCTURE OF THIS BOOK

Chapter 1 has provided an introduction to the scientific definitions used and the problems that are addressed in this book.

Chapter 2 summarizes the history of rules about consanguineous marriages according to the Roman Catholic Church, how they varied over the centuries, and historical reasons for the changes.

Chapters 3 and 4 are dedicated to an analysis started long ago, but never completed or summarized: the calculation of expected frequencies of consanguinities if marriage is random or, more precisely, is unaffected by the consanguinity degree, and the comparison between observed and expected frequencies. Chapter 3 contains a statistical analysis showing the importance of age factors affecting the frequencies of consanguineous marriages, especially differences of frequencies between degrees that differ for the number of generations leading to spouses in the two branches, husband and wife. There is a basic tendency for husbands to be older than wives in all marriages. We analyze whether the general distribution of ages of husbands and wives can explain every age effect. Statistical analysis of the data shows that the effect of age is strong and can be explained to a large extent by the general correlation for age at all marriages, but two other causes of variation must be taken in consideration. The more important one is differences in movements at marriage between husband and wife, which we call sex-differential migration. A third, minor effect we observed is a probable tendency to maintain family ties by mothers of consanguineous mates. This work was done especially for the cited three dioceses of northern Emilia (Parma, Piacenza, and Reggio Emilia), but there are some extensions to other parts of Italy.

In chapter 4 we describe simple theories to calculate the expected frequency (i.e., the probability) of consanguineous marriages, taking account of the major factors of age and sex differential migration. On their basis, we estimate how observed consanguinity deviates from random expectation, or, in other words, if a real avoidance of consanguineous marriages is supported by the need to require dispensation.

Chapters 5 and 6 are dedicated to the study of the relationship between drift and inbreeding, on the basis of genetic research done in the Parma valley. Chapter 5 describes in detail a genetic study of blood groups of the villages of the Parma valley. Its conclusions were published briefly in a *Scientific American* article (Cavalli-Sforza 1969), but the full details of the data and their analysis were never published before. Chapter 5 evaluates drift from the blood group data and also from surname data obtained from consanguineous marriages, as well as consanguinity and inbreeding data and coefficients, and compares them for the whole valley.

Chapter 6 gives, for the last three centuries, migration and other demographic data relevant to drift for 22 villages of the upper section of the Parma valley, which show the greatest drift. The analysis and comparison of observed and expected drift and consanguinity or inbreeding is made here with a computer simulation of the upper Parma valley population, made up of about 5000 individuals from the 22 villages.

Chapter 7 is dedicated to regions presenting special problems and opportunities. These include the two major islands, Sardinia and Sicily, and some other special isolates. Islands deserved special treatment, since their culture was somewhat different from that of the mainland, and it was necessary to examine the bishopric archives in detail, because not all these data are found at the Vatican archives. In Sardinia it was possible to collect data for all bishoprics over long periods. In Sicily, only first-cousin and closer consanguinities were available in the Vatican. Here a peculiarity of major interest is the very high frequency of uncle–niece matings. There are a great number of bishoprics in Sicily and not all could be examined, but trends are similar in all those examined. Chapter 7 also studies in detail a few isolates: the Aeolian Islands north of Sicily, the most important of which is Lipari.

Chapter 8 summarizes studies based on linking records of individual consanguinity with those of disease and physical traits.

Chapter 9 analyzes correlations between the frequencies of major consanguinity degrees, especially first and second cousins, and the $\alpha$ average inbreeding coefficient in all of Italy, obtained from the Vatican archives and other sources, and available data of demography, economy, and social characteristics studied by the Italian Istituto Centrale di Statistica.

Chapter 10 examines the concept of "deme" and estimates drift in all Italian communes by an analysis of surnames.

Chapter 11 summarizes the major conclusions.

The frequencies of various degrees of consanguineous marriages (percentages of total marriages) in space (92 provinces) and time (1911–1964, in 5-year periods) for the whole country, based on the Vatican records, can be found on the Internet at www.igm.cnr.it/Zei/ freqcons.html.

# CHAPTER 2

# Customs and Legislation Affecting Consanguineous Marriages, with Special Attention to the Catholic Church

## 2.1 EARLY AND MEDIEVAL CHRISTIAN TRADITION

The Catholic Church has taken a position on the issue of consanguineous marriages since its origin. Laws against marriage between relatives existed in the Jewish and Roman tradition. In the Bible, unions between close relatives (brother and sister and brother and half-sister) were severely punished: a public execution is mentioned in Deut. 20:17. Different types of forbidden matings are given by the Bible: a man could not marry his mother (Lev. 18:7), his sister (Lev. 18:12, 20:19), his father's sister (Lev. 18:12, 20:19), his mother's sister (Lev. 18:13, 20:19), or his son's daughter (Lev. 18:10). A man could marry his niece, a type of mating relatively common among Jews (Burrows 1938, Neufeld 1944, Paglino 1952), some Indian groups, and Mediterranean populations (Sicily, this book).

Roman law forbade all unions among people linked in a direct way within the seventh step (see later), with the apparent aim of favoring outbreeding. Livius reports that the patrician Celis was the first to break away from the old rules and took as a bride a relative closer than the seventh step. Later the prohibition was relaxed to the point of allowing marriages between first cousins. Theodosius in A.D. 384 again forbade this kind of marriage. We do not know the text of this Roman law but its general meaning is conveyed in a letter from S. Ambrosius (*No. 60, "Ad Paternum," no. 8*):

Nam Theodosius Imperator etiam patrueles fratres et consobrinos vetuit inter se coniugii convenire nomine et severissimam poenem statuit, si quis temerari ausus esset fratrum pia pignora; et tamen illi invicem sibi aequales sunt: tantummodo quia propinquitatis necessitudine et fraternae societatis ligantur vinculo, pietatis eos voluit debere, quod nati sunt.

[Emperor Theodosius forbade the marriage between brothers' and sisters' children and established a severe punishment for those who would dare to contaminate the innocent children of sibs, but still these are of the same age. He established that they respect the affectionate relationship set by their kinship only because they are relatives bound by a link similar to that of sibs that exists for the very fact of having been born.]

Marriages with a man's brother's daughter had always been forbidden. Emperor Claudius obtained the Roman Senate's pronunciation that disposed of the prohibition so that he would be permitted to marry Agrippina, daughter of his brother Germanicus. The same Emperor later had the Senate introduce a prohibition against marriages between uncle and niece in the case that involved a sister's daughter (Bonfante 1925).

Arcadius (A.D. 400) made marriages between relatives legal again in the eastern Roman Empire, but the new provision never became a law in the western Roman Empire. The western emperor Honorius in A.D. 409 actually forbade marriages among cousins but made the punishment lighter and introduced the possibility of a dispensation by the emperor for special cases (Esmein 1933).

The Justinian codex reiterated the prohibitions (*Institutionum, lib.1; De Nuptiis, tit. X, no.1*):

Fratris vel sororis filiam uxorem ducere non licet: sed nec nepotem fratris vel sororis qui ducere potest, quamvis quarto gradu sint. Cuius enim filiam uxorem ducere non licet, neque eius nepotem permittitur.

[A man may not marry the daughter of a brother, or a sister, nor the granddaughter, although she is in the fourth degree. For when we may not marry the daughter of any person, neither may we marry the granddaughter.]

The Catholic Church initially followed the Roman and Jewish regulations, but introduced further restrictions in the fourth century, although consanguineous marriages were regulated differently in different places and often the rules were not clear (Joyce 1948). On top of the lack of coherent rules generated by the disagreement among decrees by different authorities such as the popes and councils, further confusion was generated by the abandonment of the Roman system of consanguinity evaluation in favor of the German system.

Initially the Church used the Roman system of consanguinity evaluation in its application of consanguinity restrictions. Later the Longobard system was adopted because of its diffusion all over Italy. The canon (c. 16, C. 55, q. 2), is attributed to various popes and embodied in a letter of Gregory III (A.D. 732), which forbids marriage among the Germans to the seventh degree of consanguinity. The adoption of the German system by the Church was gradual. It was in general use by the beginning of the second millennium. It was considered directly connected with the natural relations of marriage, and Alexander II (1061–1073) treated it as a peculiarly ecclesiastical law (c. 2, C. 35, q. 5) and threatened severely all advocates of a return to the Roman, or civil, calculation. The German system excludes levels of consanguinity accepted by the Roman one, interdicting marriages between people consanguineous to the seventh degree (sixth cousins); therefore, this decision made the Church's consanguineous marriage regulations stricter. This was desirable to encourage the mixing of German and Roman populations (Penot 1902). Opposition to this switch was apparently strong in the canonical law schools like the Ravenna one. In fact, the adoption of stricter consanguinity marriage regulations made finding a mate almost impossible in a situation of increasing limitations to people's mobility. This situation also made it almost impossible to be sure that any marriage celebrated did not violate such strict consanguinity rules. There was consensus that the Levitical prohibitions were valid and the Church did not have the power to dispense from them, but it had the power to dispense from the laws that it itself generated.

A solution to the problem came with the IV Lateran Council (1215). Pope Innocent reduced the consanguinity restriction from the VII to the IV grade (third cousins):

> Quoniam in ulterioribus gradibus iam non (potuit) absque gravi dispendio huiusmodi prohibitio generaliter observari.

> [Since, in general, for further grades the prohibition could not be followed.]

Duns Scotus (1308) gave a new direction to the debate on consanguinity dispensations. He maintained that Levitical prohibitions belong to the the Old Testament and Jesus never stated that they were still binding rules under the New Testament. The basis for their validity had to be found not in the Old Testament but elsewhere:

> Sed unde est quod propinquitas talis vel talis simpliciter impedit? Respondeo in lege Evangelica non invenitur a Christo alia prohibitio ultra prohibitionem legis naturae: nec etiam explicite confirmavit prohibitionem super hoc factum in lege mosaica. Sed Ecclesia illegitimavit personas aliquando in gradu remotiori postea in quartu gradu.

> [But where is the reason why a certain level of relatedness is simply forbidden? I answer that in the law of the Gospel no other prohibition is found than the prohibitions of natural law: nor the prohibitions on this matter found in Moses' law were confirmed. But the Church sometimes made illegitimate the marriage between people related less than the fourth degree.] (Scotus lib. IV)

In summary, a marriage between relatives connected in a direct line is forbidden by the law of nature, but all other limits have been set by the Church's authority (Waddingens 1891, Fleury 1933).

Scotus' position spread in the Church's schools. In 1418 Pope Martin V, answering a request by the Count of Foix to be granted permission to marry his wife's sister, refers explicitly to the opinion of scholars of theology and canonical law that it was in the pope's power to grant dispensations from Levitical prohibitions. Although popes had apparently never used this power, in the fifteenth century they started granting such dispensations.

The final prevailing position among Catholic theologians was that natural law, as we know it through reason, forbids only marriages between parents and children and between sibs. It indicates that marriages between close relatives are improper and disorderly (and the Levitical relationships are an example) and that the latter marriages require a justification of relevance comparable with the degree of relatedness, and, lastly, that a society in which such marriages would be freely celebrated would risk a lowering of its moral standards (Sanchez 1607).

The Trento Council set these opinions in a Canon on Marriages. The third Canon recites, "If anybody will say that only the levels of consanguinity indicated by the Levitic can forbid that a marriage be celebrated or that, if celebrated, such a marriage is null; and that the Church cannot grant a dispensation for some of these degree of consanguinity or decree that other degrees, beside these, forbid or make null the marriage, anathema" (*Sacros. Conc. Trid.,* Sessio XXIV, 11 November 1563).

It is interesting to note how the Church often accepted the manners and institutions of newly converted peoples. The Catholic Encyclopedia (1955) reports numerous examples. Gregory I (590–604) granted to the newly converted Anglo-Saxons restriction of the impediment to the fourth degree of consanguinity (c. 20, C. 35, qq. 2, 3); Paul III restricted it to the second degree for American Indians (Zitelli, Apparat. Jur. Eccl., 405), and also for natives of the Philippines. Benedict XIV (Letter "Æstas Anni," 11 October, 1757) states that the Roman pontiffs have never granted dispensation from the first degree of collateral consanguinity (brothers and sisters). For converted infidels it is recognized that the Church does not insist upon annulment of marriages beyond this first degree of consanguinity. (For further details of the history of ecclesiastical legislation concerning this impediment, see Esmein [1933]).

Consanguinity regulations have been organized in the Canonical Law (Corpus Juris Canonici) of 1917. The consanguinity impediment was reduced from the IV to the III degree. In the new Canonical Law of 1983 restriction to consanguineous marriages was reduced to the second degree, also ignoring cases of multiple consanguinity.

## 2.2 TRADITIONAL METHODS OF CONSANGUINITY EVALUATION: THE ROMAN AND THE GERMAN METHODS

It may be of interest to briefly review the two different ways of attributing a level (grade) of consanguinity to the relationship between two relatives according to the two methods (Roman and German) that have been referred to previously (see also figure 1.1). In the Roman method (*computatio Romana civilis*) the consanguinity grade is given by the number of times the generation process (*processus generationis*) has taken place: from one relative the number of generations are counted until the common ancestor is reached and then the number of generations that link the common ancestor to the other relative are added (Joyce 1948). The grade of consanguinity is given by the total number of generations (*processi generationis*) that link the two relatives. According to this method the relationship between a brother and a sister has a second-grade consanguinity, between uncle and niece third grade, first cousins fourth grade, second cousins sixth grade, and so on. Conveniently, the number of people that link the two relatives (omitting the common ancestor) gives us the degree of consanguinity. The Latin language has specific names for relatives up to the sixth grade. In Roman law inheritance rights were limited to the sixth grade of consanguinity. The Church adopted the practical definition of relative as the person who was allowed to inherit property. As a consequence, marriage was considered consanguineous and forbidden within the same grade of relatedness. The Roman Church extended the restriction to the seventh grade to avoid marriages between people more related than third cousins.

Germans, Francs, Longobards, and other northern populations did know Roman traditions and measured consanguinity in a different way as the number of generations that separated the two relatives from the common ancestor. According to this method (*computatio Germanica*) the relationship between a brother and a sister has a first-grade consanguinity, first cousins second grade, second cousins third grade, and so on. To obtain the consanguinity grade it was enough to count the number of people between one relative and the common ancestor, omitting the common ancestor. For relatives connected by an uneven number of people to a common ancestor both lines had to be mentioned. An uncle and niece had a second-grade combined with a first-grade consanguinity. The general rule was

Tot gradibus collaterales distant inter se quot uterque (remotior) distat a stipite communi (dempto stipite).

[Relatives are linked by a grade of consanguinity equal to the distance that both (or the more distant) have from the common ancestor (the latter not considered).]

It is now clear why the adoption of the German method would extend the impediment of consanguinity.

In the German method the structure of the human body was used to compute consanguinity grades: the head was the common ancestor, the shoulder was the first grade, the elbow was the second, the wrist was the third, and the finger joints were examples of up to the seventh grade.

## 2.3 JUSTIFICATIONS OF THE DISPENSATION REQUEST

A curiosity that opens up an interesting window on the times in which the events analyzed in this book took place is the formulae suggested for use in preparing an application for the dispensation from consanguinity impediments (*Formulae Apostolicae Datariae pro matrimonialibus dispensationibus jussu*, 1901). Here is the list divided into two sets of causes, which suffice, either alone or concurrently with others. This list of causes is by no means exhaustive; the Holy See, in granting a dispensation, will consider any weighty circumstances that render the dispensation really justifiable.

1. Limited size of the locality (*Restrictio loci*). It is considered absolute when the village has less than 1500 inhabitants (or less than 300 families) and the proportion of relatives is so large as to make it impossible for the applicant woman to find a husband of equal social condition who is unrelated and it is too hard for her to move to another village. Even a more populous location is considered of limited size when suitable unrelated men are scarce due to a war or to the fact

that the woman belongs to a Catholic minority within a non-Catholic community.

2. Superadult age of the applicant woman. In this context a woman is considered of superadult age if she is between 24 and 50 and has not yet found a husband.

3. Lack or small size of the applicant's dower. The justification can be invoked when the relative she intends to marry is willing to provide a dowry.

4. Poverty of an applicant widow, burdened by a large number of children.

5. Favoring peace (*Bonum pacis*). Valid when a consanguineous marriage ends serious litigation, hatred, and animosity between families.

6. Orphan condition (both parents) of the applicant woman.

7. Deformity, physical imperfections, and sickness of the applicant woman.

8. Validation of a wedding celebrated in good faith ignoring the need of a church dispensation

9. Imminent wedding whose cancellation would procure severe moral and economic damages to the applicants (*Quando omnia sunt parata*).

10. Mutual aid in the case of a marriage between older people (above 50).

11. Wedding already announced whose cancellation would generate derogatory suspicions.

12. Favoring the well being of children (*Bonum prolis*). To be used when one or both applicants are widowed and the marriage would make possible to support, educate, and assist orphan minors.

13. Mutual familiarity of the applicants.

14. Infamant suspicions on the applicant woman generated by her engagement that could severely affect her future chances of getting married.

15. Suspicious cohabitation in the same house that cannot be easily interrupted (*suspicio copulae*).

16. Determination of the two engaged people to pursue their intention of marrying.

17. Danger of an incestuous concubinal relationship.

18. Risk of the celebration of a civil wedding, to be used when a civil dispensation has been obtained or requested or when a civil wedding has already been celebrated.
19. Pregnancy, need to make the children legitimate.
20. Removal of a public scandal or of a well-known concubinal relationship.
21. Loss of virginity by the bride with a person different than the groom.
22. The applicant woman's condition of having been born as an illegitimate child.
23. Eloping. To be used only if the woman has been returned to a safe place and put in a condition of freely consenting to the marriage.

A further set of situations could be used either to support a dispensation request for more distant relatives or to make a request of dispensation for close relatives stronger:

1. Solution of inheritance fights.
2. Need by the applicant man of that specific woman for his special necessities: as an aid in running the family business or in nursing his bad health.
3. Good reputation of both applicants.
4. Special convenience of the marriage for special needs.
5. The well being of parents in need of help (*Bonum parentum*).

This is a suggested list of "formulae" that were followed in most applications but that was in no way exhaustive. In fact, some reveal quite a bit of fantasy on the part of the priest that most likely helped in the preparation of the application. So both applicants being gypsy, one of them being feebleminded or blind or having lost a limb in war, or helping a brother of one of the applicants avoid being drafted into the army are all justifications invoked for granting the dispensation. Even a reference to the participation of an ancestor to the crusades has been found.

The formulae that were used by most applications suggested the possibility of a quantitative study (Boiardi 1961). More than 1500 dispensations from the Piacenza diocese were examined from the period 1928–1958. Following the national trend, most marriages requir-

TABLE 2.1. Diocese of Piacenza, 1928–1958: the most frequent motivations quoted in consanguinity dispensation requests, distributed by altitude of applicants' village of residence

| | Mountain | | Hills | | Plain | | City | | |
|---|---|---|---|---|---|---|---|---|---|
| Motivation | No. | % | No. | % | No. | % | No. | % | Total |
| 1 Small village size | 555 | 41.95 | 155 | 37.71 | 101 | 28.13 | 11 | 7.80 | 822 |
| 2 Bride's older age | 291 | 21.99 | 99 | 24.09 | 130 | 36.21 | 50 | 35.46 | 570 |
| 3 Lack of dowry | 129 | 9.75 | 37 | 9.00 | 20 | 5.57 | 15 | 10.64 | 201 |
| 13 Familiarity | 121 | 9.14 | 36 | 9.76 | 4 | 1.11 | 22 | 15.60 | 183 |
| Total motivations | 1,323 | | 411 | | 359 | | 141 | | 2,234 |

ing consanguinity dispensations were celebrated in the mountains (59%) with a clear decreasing trend moving down to the hills (18%) and in the plain (16%). The rest of the weddings (7%) were celebrated in the city or outside the dioceses.

The applications invoked more than 2,200 motives, a clear indication that a good proportion of applicants indicated more than one motivation for their request. Clearly, there was a range of attitudes toward the granting process: from the worried ones that listed more than one justification to some that do not indicate any motivation at all. The prevalence of a bureaucratic attitude can be inferred from the fact that the first three motivations of the formulary (village size, older age of the applicant woman, and lack of dowry) are also the most often quoted: alone or in combination with other motivations they appear in more than 70% of the considered marriages. Familiarity between the applicants seems to be used to summarize the moral concerns that might justify the application and is the only one of the other causes used in a relevant number of cases (12%). The picture does not change much if we consider the motivation by altitude (table 2.1). Obviously, the smallness of the village is not invoked in the city: socioeconomic (lack of dowry) and moral (familiarity) considerations seem to prevail.

# Demographic Factors Affecting the Frequencies of Consanguineous Marriage—A Study in Northern Emilia

## 3.1 NATURE AND INTEREST OF THE PROBLEM

The subject of this and the following chapter is the prediction of the frequency of consanguineous marriages of given type and degree, based on the assumption that marrying a consanguineous person takes place randomly. The comparison of these expected frequencies with observed ones allows us to detect possible deviations from randomness, due to preference for or avoidance of consanguineous marriages. The calculation of expected frequencies requires knowledge of certain quantities, mostly of demographic nature. We have been able to secure such knowledge from several sources in or around the Parma diocese, located in northern Emilia, in the middle of the Po valley in northern Italy. These are also the regions where we have the most detailed data on consanguineous marriages, but we will also extend the analysis and the comparison of observed and predicted frequencies of consanguineous marriages to other regions where partial data have been obtained.

Our interest in predicting the frequency of consanguineous unions is that it allows us to understand the factors that affect such marriages. Observation has shown that their frequencies can vary greatly according to population and from place to place and time to time. Such variation may contain the keys for understanding the determining factors. The frequencies of such unions, on the other hand, are essential for predicting the inbreeding of the population, a quantity that is part of "population structure." Dahlberg suggested many years ago that, under simple hypotheses, the frequency of consanguineous unions might allow the calculation of the *isolate size*, or the number of individuals among whom a mate is chosen. This is also a parameter of population structure. In the following section we present a simplified description of the main aspects of this important part of the study of evolution.

## 3.2 RELATIONS TO POPULATION STRUCTURE

Population structure is as an evolutionary factor involving the chance that two individuals of a specific population will mate with each other. This includes a number of aspects related to the evolving population, completely separate from the two major ones: mutation and natural selection. When the whole species is considered, it may, for simplicity, be reduced to one quantity: the overall population size $N$. But the simultaneous presence of more than one generation in any human population makes it necessary to reduce $N$ from the total census size to the number of individuals effectively reproducing, $N_e$, as explained in chapter 8 of Cavalli-Sforza and Bodmer (1971, 1999). The effective population size of the species determines the amount of *random genetic drift* (or simply *drift*), that is, the effect of stochastic fluctuations of gene frequencies in time for the whole species, provided the species is not subdivided into essentially noncommunicating sections, which will then tend to behave independently. Clearly, the change of $N_e$ with time is important, in spite of the rarity of research related to it.

This is only the tip of the iceberg of population structure, especially if we are interested in the behavior in time and space of the subsections into which it is subdivided. Much more detail is necessary for defining the subdivision of the species into populations showing a certain degree of independence from each other in mating. The effective sizes of such populations are of importance for the problem of the frequency of consanguineous matings, because they define the communities within which a mate is usually chosen. They thus correspond to the "isolates" of which Dahlberg was trying to estimate the sizes. Limiting estimation to the size of the "isolates" is not enough, however, because human populations are very rarely completely isolated. The amount of intermigration between populations is usually high, especially if the isolate is really small. In addition, even small human populations tend to have some internal social structure that further complicates the picture.

The genetic variation (of gene frequencies) between neighboring isolates is, therefore, determined by their population size, the genetic exchange among them ("migration"), and further possible factors of internal social structure determining mating preferences. The degree of inbreeding and the frequencies of consanguineous matings depend on these quantities. They are usually difficult to evaluate or predict, so there is an interest in each of them for the knowledge they can supply on the other, related ones.

Dahlberg suggested (1926, 1948) that in isolated communities, for example, in small remote villages, the relative frequencies of consanguineous marriages is determined by the size of the "isolate" and can be used to measure it. If people marry exclusively in the village, and marriages occur randomly, the expected frequency $\Phi$ of a given type of consanguineous marriage (e.g., between first cousins) is given by the ratio of the number $C$ of consanguineous individuals of that type, and the number $N$ of individuals in the village. For example, if $C = 4$ for first cousins, then $\Phi = 4/N$ and $N = 4/\Phi$.

This hypothesis may explain the well-known fact that the frequency of consanguineous marriages tends to increase in areas with small and isolated villages, where $N$ is small and migration infrequent, as is true of most mountainous areas, small islands, and certain social groups, like royalty. It may also explain the decrease of consanguineous marriages and the increase in the geographic distance between birthplaces of spouses, observed especially in isolated areas, and starting in the nineteenth century, which Dahlberg called the "breakdown of isolates." It is most probably due to the increased migration made possible by the increasing communication (railways, new roads, new mechanical means of transportation), which is typical of modern times.

The concept of "isolate" has been replaced more recently by the term *deme*, meaning a population whose members mate randomly. It is the equivalent of the older term "Mendelian population." The concept and its application are further discussed in chapter 10.

The calculation of isolate size might have made Dahlberg's approach useful in population genetics, were it not that the approach, as it was formulated, had serious approximations that made it unusable. Morton (1955) showed that the isolate size calculated by Dahlberg's formulas varies by a factor of 1 to 1000 or more when it is calculated from first cousins, second cousins, 1½ cousins, uncle–niece, and aunt–nephew, in this order.

There are further difficulties for a precise analysis. For instance, there is a significant difference in the frequencies of the four existing types of first cousins, depending on the relationships of the parents of the consanguineous mates. These can be two brothers, one brother and one sister, or two sisters. There are two brother–sister types of first cousins, depending on whether the brother is the father of the consanguineous husband or of the wife. We will call the four types bro–bro, bro–sis, sis–bro, sis–sis.

It is clear that at least some of these inconsistencies must depend on the strong correlation of spouses' ages, which must make certain marriages less or more likely to occur. The very high isolate size calculated from aunt–nephew or uncle–niece marriages may depend, for instance, on the big age difference between the prospective mates. Clearly, there is an average difference of one generation between an uncle and a niece or an aunt and a nephew, and this simple fact makes these marriages unlikely to occur. Moreover, since husbands tend to be older than wives in most populations, an uncle–niece marriage may be more likely than an aunt–nephew, as is indeed observed. But it is also true that husband and wife tend to be of similar age, and therefore a marriage between first cousins is more likely, a priori, than one between an uncle and a niece or an aunt and a nephew.

Age of mates explains a good fraction of the differences in frequencies of the various types and degrees of consanguineous marriages, but other factors exist that have a major influence (Barrai et al. 1962). One of these was the differential migration of the two sexes. In a highly agricultural area, such as the rural ones of Italy and most of Europe until the beginning of last century, most migration was a consequence of marriage. Men usually inherit the land, and women who do not marry in the same village in which they are born usually join their husbands; they therefore migrate more often than men. Industrialization has altered these rules, but only recently. It also became apparent that the number of consanguineous individuals of given degree was underestimated in the approach by Dahlberg. There is an interaction between age and migration effects, making it preferable to test them jointly for greater precision. For many purposes, however, it is sufficient to use a simpler procedure, especially because the frequency of consanguineous marriages varies with space and time enough to make it impossible to expect a perfect fit of expected frequencies to observed ones.

This analysis is limited to simple consanguinities. There are records of multiple consanguinities (in which the mates are joined by several loops of relationships of various depth), but they are rare. The simplest cases are double first cousins (when the parents of the consanguineous mates are two brothers who married two sisters, or one brother and sister who married a sister and brother) who are as closely related as uncle–niece or aunt–nephew; other cases are first cousins and second cousins, and so on. In chapter 6 we show that when there is more than one consanguineous loop in a pedigree, the more remote ones have often been neglected in the consanguinity records, and this is a cause of several approximations.

## 3.3 NUMBER OF SIBS, DISTRIBUTION OF FAMILY SIZES, AND OBSERVED ABUNDANCE OF RELATIVES

In his attempt at predicting the frequency of consanguineous marriages, Dahlberg estimated the number of relatives of a given degree with a formula that does not take into account the variation of number of children per family. In general, the ancestor common to the *propositus* (an individual on whom we fix our attention) and his relatives of a specified type is the last common one, and the relationship of the propositus and his consanguineous relatives is established by two sibs who are children of the last common ancestor or ancestors. The calculation of the number of sibs therefore has a central part in the calculation of the number of relatives of a given degree. Dahlberg took the number of sibs of an individual as equal to $n - 1$, where $n$

is the size of the family. He assumed that the size of the family varies only "little" (more precisely, does not vary at all) from one family to the other. But in reality the variation of the number of children per family is not at all negligible: ignoring it introduced a serious underestimate of the number of sibs, by a factor of 2 or usually more (Frota-Pessoa 1957, Mainardi et al. 1962, Cavalli-Sforza et al. 1966).

Each child from a family with $n$ children has $s = n - 1$ sibs, but the family must be counted $n$ times because it has $n$ children. The number of sibs that counts is that of those who survive to marry and actually marry (more generally, who continue the species; that is, the "reproducing" children). If there was no variation of $n$ from family to family, as assumed by Dahlberg, the average number of sibs would be exactly $n - 1$. But if $n$ varies from family to family, the frequency distribution of the number of sibs is given by

$$F(s) = \frac{n \times f(n)}{\Sigma n \times f(n)}$$

where $f(n)$ is the frequency of families with $n$ children, $F(s)$ that of sibs, $s$ ($= n - 1$), and $\Sigma$ is taken for $n = 1$ to infinity. This distribution has an average:

$$\text{ms} = \text{mean number of sibs} = \frac{V}{n} + n - 1 \qquad (3.1)$$

where $n$ is the mean number of reproducing children, and $V$ is their variance (Cavalli-Sforza et al. 1966). When this number follows the Poisson distribution and, therefore, $V = n$, then the average number of sibs is equal to the average number of children per family. As mentioned before, the latter is not to be taken equal to the total number of children, but just as that of the reproducing ones. In a demographically stationary population the number of reproducing children is on average 2 per family, if everybody marries, and, therefore, the number of sibs is also 2 per family, if the number of children is Poisson distributed. In this case the average number of sibs is twice as great as that assumed by Dahlberg, who assumed no variation of number of children per family.

The number of reproducing children per family is clearly of importance for our calculations. There are two possible complications. The first is when a population is not demographically stationary. Then the average number of reproducing children per family is not 2, but is 2 multiplied by the net growth rate per generation (the net growth rate per year powered to the number of years of the average generation time). The growth rate is less than one, when the population is decreasing, and greater than one if it is increasing. A second complication depends on the variance of the number of children per family; if this is not equal to the mean, then the average number of sibs is not equal to the average number of children. It will be greater or smaller, depending on whether the variance of the number of children is greater or smaller than mean number of children. Formula 3.1 gives its expected value.

Strictly speaking, there is no need to know the shape of the distribution of the number of children: formula 3.1 shows that knowledge of its mean and variance is sufficient for our purposes. In other circumstances it may be useful to know the shape of the distribution. Demographers are familiar with the use of the negative binomial, which often fits observed distributions of the number of children per family, though imperfectly. It is the distribution expected if the fertility varies from family to family according to a gamma distribution (also called chi square or Pearson type III), and the number of children born per family varies randomly according to the Poisson distribution with mean equal to the fertility of the individual family. Demographers (e.g., Brass, 1958), usually fit the negative binomial distribution to the "total" numbers of children per family and such results are not useful for our purposes.

There is less experience of fitting a theoretical distribution to the number of "reproducing children" (i.e., children who have progeny). The geometric distribution (a special case of the negative binomial, with $k = 0$) was found approximately useful in some cases. Distributions of the numbers of reproducing children were calculated in the Parma valley from reconstructions of pedigrees since the beginning of parish books (starting in the eighteenth century). Counts generally made were those of the number of children who married per married couple. The first reconstruction was made by hand in one mountain village (Cavalli-Sforza 1957), then over a wider area around it by computer (Skolnick 1974). Incompleteness of the parish records has caused the mean number of reproducing children to be less than 2, although the population of the mountainous part was on average stationary over two centuries, though decreasing over the eighteenth century and increasing in the nineteenth. The observed distributions are not fitted perfectly by any theoretical distribution tried; strictly speaking imperfect fits, but reasonably adequate as a first approximation, were obtained with the negative binomial and also with the geometric (Skolnick et al. 1971, Skolnick 1974). The approximation of the geometric can be easily appreciated visually by plotting the logarithms of the number of families with $n$ children versus $n$. A straight, descending line is expected.

A correction for the proportion $b$ of families with zero children is necessary using a "truncated" or "dudded" geometric distribution (Karlin 1966). If the average number of children is $m$, then from the truncated geometric the average number of sibs is $m/1 - b$. For a demographically stationary population with $m = 2$, the average number of sibs is greater than 2. But, in principle, the approximation introduced by $s = 2$ is modest.

The observed distributions of number of reproducing children from the Parma valley gave rise to curves that are not too different from straight lines, but showed some waviness. No substantial improvement in the goodness of fit was obtained by plotting the extra parameter needed for a truncated negative binomial (Skolnick et al. 1971).

Empirical estimates of numbers of close relatives have been made much more frequently on living people than on reproducing ones. An example given in Table 3.1 shows counts of living relatives (uncles and aunts, first cousins) on 423 families living (in 1958) in the Parma valley. Families were not all complete at the time of sampling.

The negative binomial distribution was fitted by calculating parameters $p$ and $k$ given in the table. The fit was reasonably good in the first four cases. The distribution of first cousins has a mean reasonably similar to that expected, which is the product of the means of the number of living children and their married uncles. But it has a variance significantly greater (99.77) than that expected (69.64); this variance is likely to be particularly sensitive to the correlation of fertility from generation to generation, due to genetic or socioeconomic causes or both (Mainardi et al. 1962).

TABLE 3.1. Survey of the number of relatives of the children in 423
families of the Parma valley

| Number of | Mean | Variance | p | k |
|---|---|---|---|---|
| Living children | 2.497 | 3.356 | 0.344 | 7.258 |
| Their uncles (total) | 8.397 | 15.520 | 0.848 | 9.898 |
| Their living uncles | 6.281 | 11.326 | 0.803 | 7.820 |
| Married uncles | 6.201 | 10.734 | 0.731 | 8.483 |
| Living first cousins | 14.105 | 99.766 | | |

Source. Mainardi et al. (1962).
Note. p and k are parameters of the fitted negative binomial.

## 3.4 CONSANGUINITY DEGREES AND OBSERVED NUMBERS OF CONSANGUINITY DISPENSATIONS IN NORTHERN EMILIA

All the consanguineous marriages in which we are interested involve indirect relatives and therefore descendants of sibs, who lived a number of generations back. We will call *consanguinity degrees* the following six major categories of consanguineous marriages. At the beginning of each line is indicated the inbreeding coefficient $F$ that belongs to each of the six categories:

$F = 1/8$      uncle–niece (UN) and aunt–nephew (AN), or 12
$F = 1/16$      first cousins, or 22
$F = 1/32$      first cousins once removed (1½), or 23;
$F = 1/64$      second cousins, or 33
$F = 1/128$      second cousins once removed (2½), or 34;
$F = 1/256$      third cousins, or 44

The first, third, and fifth degrees, all of which are called *uneven cousins* (while the second, fourth and sixth are called *even cousins*), are made of two categories (1) with the husband and (2) with the wife, on the shorter branch of the pedigree. For the first degree these correspond to UN and AN, respectively. The last two categories are available only for older periods, since Roman Catholic rules stopped requiring dispensations for them after 1917.

Our best data come from consanguinity dispensations from the archives of the diocese of Parma, from 1851 to 1957. A summary is

TABLE 3.2. Number of consanguinity dispensations in the diocese of Parma, subdivided into two major periods, by consanguinity degree (see text) and by type of cousins on the basis of genealogy data available for each dispensation

| F | Type of cousins | | | | |
|---|---|---|---|---|---|
| | Full | Half | Multiple | Without tree | Total |
| **1851–1917** | | | | | |
| 1/8 | 31 | 0 | 11 | 3 | 45 |
| 1/16 | 740 | 3 | 40 | 52 | 835 |
| 1/32 | 247 | 2 | 50 | 24 | 323 |
| 1/64 | 1,072 | 1 | 60 | 165 | 1,298 |
| 1/128 | 423 | 0 | 29 | 60 | 512 |
| 1/256 | 569 | 0 | 0 | 61 | 630 |
| Total | 3,082 | 6 | 190 | 365 | 3,643 |
| **1918–1957** | | | | | |
| 1/8 | 14 | 0 | 2 | 0 | 16 |
| 1/16 | 690 | 0 | 13 | 12 | 715 |
| 1/32 | 241 | 4 | 22 | 10 | 277 |
| 1/64 | 754 | 5 | 1 | 78 | 838 |
| Total | 1,699 | 9 | 38 | 100 | 1,846 |

shown in table 3.2, subdivided into the period for which data for all six degrees are available (1851–1917) and those for which only the first four degrees have been collected (1918–1957). In this table the 5489 Parma marriages are subdivided into four categories specifying the general type of consanguinity:

1. *Full cousin marriages*, which are the greatest majority (84.6 and 92% in the two periods).

2. *Half-cousin marriages* in which the two sibs leading the pedigree are from two different marriages of a common ancestor (half-sibs). These are rare, given that divorce was not possible in Italy until recently, but there are so few of them (15 out of 5489) that perhaps they were not always recorded as half-cousin marriages. Their inbreeding coefficient is 1/2 of that indicated in the row.

3. *Multiple consanguinities* (from 5.2 to 2.0%), which would have an F somewhat higher than that indicated in the corresponding row. When the consanguinities occurring within the

same pedigree are different, the marriage has been assigned to the higher consanguinity; for example, a marriage between two spouses who are both first cousins and second cousins would have an $F$ equal to the sum of $1/16$ and $1/64$ ($5/64$) and was listed under first cousins. At most, multiple consanguinity involves doubling of the $F$ value indicated in the row (e.g., double first cousins have $F = 1/8$).

4. *Without a tree* (10 and 5.4%) for which one cannot distinguish the pedigree type.

Table 3.3 shows the data of the full cousin marriages of the Diocese of Parma subdivided by degree and into 20-year periods. It also shows the estimated number of marriages in each period, allowing a calculation of the relative frequencies of each degree and of consanguineous marriages on the total. The diocese and the province do not completely overlap. The total numbers of marriages had to be estimated from the data on the province of Parma, taking account of the proportion of inhabitants belonging to the diocese in each commune of the province, and of the percentage of marriages in proportion to inhabitants in the various periods. It was assumed that all marriages were celebrated according to the Catholic rite; the relative frequencies of consanguineous marriages in the last row of the table are therefore slightly underestimated.

There has been some variation in the proportion of consanguineous marriages over time, from a minimum of 2.56% to a maximum of 3.46% (excluding the three later periods in which 2½ and third cousins are not fully available). There is a slight tendency to a frequency increase of the close consanguinities in the first three periods, while the most remote consanguinity degree decreases in frequency. In the second half there is a tendency to a general decrease; the last two degrees, being no longer recorded, have zero frequency. The causes of most of this temporal variation are complex and are discussed in chapter 9. But there are important signs of consistency.

Data almost as good are available for the same periods from the diocese of Piacenza, and slightly less detailed ones are available from the diocese of Reggio Emilia. The three dioceses are adjacent and form the northern part of the Italian region known since Roman times as Emilia, which extends along the Po valley to the river mouth.

TABLE 3.3. Dispensations requested in the diocese of Parma from 1851 to 1957, subdivided by 20-year periods and by the six degrees of consanguinity, and estimates of the total numbers of marriages occurring in the same period

| Consanguineous marriages | 1851– 1870 | 1871– 1890 | 1891– 1910 | 1911– 1930 | 1931– 1950 | 1951– 1957 | Total |
|---|---|---|---|---|---|---|---|
| Uncle–niece Aunt–nephew | 6 | 11 | 12 | 9 | 5 | 2 | 45 |
| First cousins | 86 | 199 | 339 | 480 | 279 | 47 | 1,430 |
| First cousins once removed | 52 | 84 | 85 | 136 | 111 | 20 | 488 |
| Second cousins | 238 | 315 | 412 | 434 | 352 | 75 | 1,826 |
| Second cousins once removed | 114 | 154 | 120 | 35 | 0 | 0 | 423 |
| Third cousins | 212 | 193 | 138 | 26 | 0 | 0 | 569 |
| Total | 708 | 956 | 1,106 | 1,120 | 747 | 144 | 4,781 |
| Total marriages | 27,660 | 30,307 | 32,000 | 37,016 | 42,859 | 14,542 | 184,384 |
| Consanguineous/ total % | 2.56 | 3.15 | 3.46 | 3.03 | 1.74 | 0.99 | 2.59 |

TABLE 3.4. Frequencies and percentages of consanguineous marriages by degree in the three dioceses of northern Emilia, cumulated for the period

| Consanguineous marriages | Parma | | Piacenza | | Reggio Emilia | | Total |
|---|---|---|---|---|---|---|---|
| | No. | % | No. | % | No. | % | |
| Uncle–niece Aunt–nephew | 45 | 0.9 | 60 | 0.6 | 25 | 0.4 | 130 |
| First cousins | 1,430 | 29.9 | 2,534 | 26.7 | 1,217 | 21.6 | 5,181 |
| First cousins once removed | 488 | 10.2 | 1,006 | 10.6 | 487 | 8.6 | 1,981 |
| Second cousins | 1,826 | 38.2 | 3,162 | 33.3 | 2,293 | 40.6 | 7,281 |
| Second cousins once removed | 423 | 8.8 | 1,123 | 11.8 | 625 | 11.1 | 2,171 |
| Third cousins | 569 | 11.9 | 1,615 | 17.0 | 1,000 | 17.7 | 3,184 |
| Total | 4,781 | | 9,500 | | 5,647 | | 19,928 |
| Total marriages | 184,384 | | 356,961 | | 209,444 | | 750,789 |

They are very similar to each other in ecological terms, including approximately equally large mountainous, hilly, and plains sections. Barrai et al. (1962) listed only a fraction of these marriages, especially because the Reggio Emilia data were limited to the last 30 years and the two periods before and after 1917 were not kept distinct.

When the three northern Emilia dioceses are compared, as in table 3.4, there is a remarkable similarity of the frequencies. The overall incidence of consanguineous marriages varies from 2.59 to 2.70% in the three dioceses. The close ecological and historical similarity of the three dioceses is clearly responsible for this result.

Let us start with a simple, general analysis of the phenomenon. The frequency of the six degrees of consanguineous marriages must be affected by the number of relatives, which increases in the simplest condition (a stationary population) as the powers of 2, as degrees become more remote, being equal to 2 for UN or AN, and 64 for third cousins. But there is no such a tendency in the observed frequencies, which show a peak for second cousins, then a decrease.

It is quite clear that the three even degrees (first, second, and third cousins) show the highest frequencies. Even cousins tend to have similar ages, more than two random members of the population, and there is a high correlation for age between husband and wife. Uneven cousins are bound to have greater age differences, there being on average one generation difference between husband and wife. UN and AN have the lowest frequencies, but the age dissimilarity is bound to decrease with increasing degree; that is, 1½ cousins have less dissimilar ages than AN or UN, and 2½ less than 1½.

The age effect, however, can hardly explain the drop of frequency in third cousins compared to second cousins. On the basis of the number of relatives, third-cousins and second cousins should be in a ratio of 4 to 1, but third cousin marriages are less frequent than second cousins marriages. The same ratio, 4, is expected for second to first cousins, while the observed ratio is not even as high as 1.5. The relative lack of third cousins might be due to the difficulties of identifying them. Priests were careful in searching for relationships, even if remote, and had the necessary instruments in parish books, which they themselves prepared and stored. If the consanguinity of prospective spouses is missed and marriage celebrated without requesting a dispensation the marriage would have to be dissolved, generating a serious problem. One cannot completely exclude that some remote consanguinities would be more easily missed than the near ones, but it would seem logical to look also for some other factor as a cause of this discrepancy. A major practical problem presenting itself to priests was the need of consulting books from other parishes for a full reconstruction of genealogies, and this may have frequently been omitted.

Another factor known to be important is migration, which was studied by Barrai et al. (1962) and Cavalli-Sforza et al. (1966). The more remote the consanguinity, the wider the dissemination of relatives from their place of common origin. Marriage tends to take place between spouses living near each other; migration of ancestors decreases the chance of marriage between more remote relatives. Cavalli-Sforza (1966) indicated the possibility that these two factors are not enough to explain the data, and that there may be a tendency at least in certain populations, in particular in Parma, to avoid close consanguineous marriages probably because of fear of genetic damage to progeny.

We plan to investigate quantitatively these factors. For this purpose it will be necessary to carry out a subtler analysis of the various types of consanguinity that can be distinguished within consanguinity degrees. We will then analyze age, migration, and other factors to predict quantitatively their effects and make a synthesis of these predictions. This will help us to see if and which possible factors have been left out.

## 3.5 PEDIGREE TYPES, PEDIGREE CODES, AND PROOFS OF THE INFLUENCE OF AGE AT MARRIAGE AND OF THE SEX OF INTERMEDIATE ANCESTORS

The six consanguinity degrees to which we limit our study in this chapter include a number of distinguishable genealogies, or *pedigrees*, depending on the sex of ancestors included in the genealogy. Distinguishable pedigrees are shown in figures 3.1–3.3, where squares and circles represent male and females ancestors, respectively.

We will call *common ancestor/s* the parents of the two sibs who

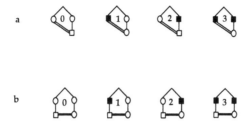

FIGURE 3.1. Pedigree types and their code numbers: (a) uncle–niece and aunt–nephew; (b) first cousins.

start the two lineages leading to the consanguineous husband and wife; and *intermediate ancestors* those found in the pedigree between the common ancestor/s and the consanguineous husband and wife.

We also state that

$i$ is the number of intermediate ancestors on the left branch;

$j$ is their number on the right branch;

$m_i$, $m_j$ are the numbers of males among intermediate ancestors in the left and right branches;

$f_i = i - m_i, f_j = j - m_j$ those of females.

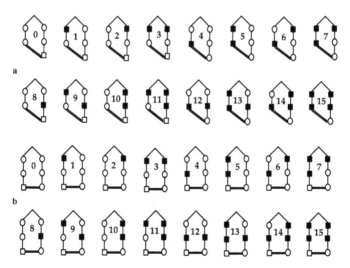

FIGURE 3.2. Pedigree types and their code numbers: (a) first cousins once removed (1½); (b) second cousins.

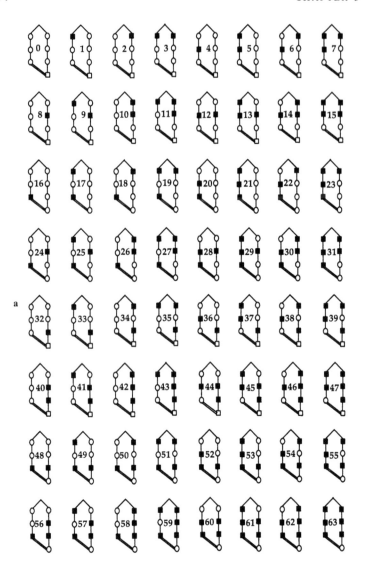

a

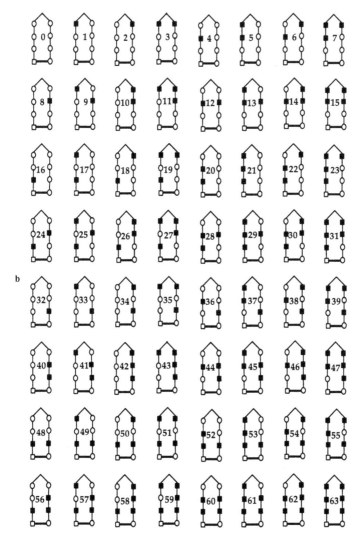

FIGURE 3.3. Pedigree types and their code numbers: (a) second cousins once removed (2½); (b) third cousins.

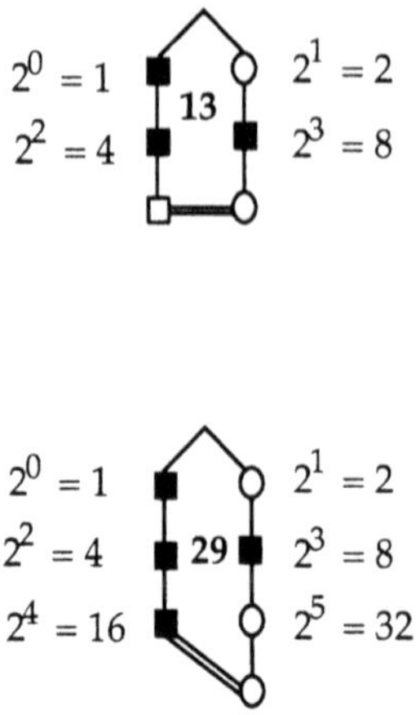

$$2^0 = 1 \quad \blacksquare \quad \bigcirc \quad 2^1 = 2$$
$$\mathbf{13}$$
$$2^2 = 4 \quad \blacksquare \quad \blacksquare \quad 2^3 = 8$$

$$2^0 = 1 \quad \blacksquare \quad \bigcirc \quad 2^1 = 2$$
$$2^2 = 4 \quad \blacksquare \; \mathbf{29} \; \blacksquare \quad 2^3 = 8$$
$$2^4 = 16 \quad \blacksquare \quad \bigcirc \quad 2^5 = 32$$

FIGURE 3.4. Two examples of calculation of pedigree code numbers.

Each pedigree is given a code number within the consanguinity degree to which it belongs; the code number is indicated in the figures inside each pedigree, and is calculated according to the procedure shown in figure 3.4 for two examples, one of second cousins and one of 2½ cousins.

The husband is written at the lower left of the pedigree for even cousins, and the shorter branch is written at the left for uneven cousins. The code number is the sum of powers of 2 for all positions in the pedigree at which there is a male. Positions counted are those of intermediate ancestors for even cousins, and include the spouse of the short branch for uneven cousins. Positions are numbered starting from the first intermediate ancestor at top left, reading from left to right and top to bottom, as shown above. One can use the pedigree code number as indicating directly the positions of males among intermediate ancestors, by decoding it as a binary number. For instance, pedigree 37 of third cousins can be decoded into $37 = 1 + 4 + 32 = 2^0 + 2^2 + 2^5$. Pedigree 37, therefore has three male intermediate ancestors in positions 0, 2, 5.

The pedigrees of simple consanguineous marriages have mostly two common ancestors, not indicated in the pedigrees. There can also be pedigrees with a single common ancestor who married twice, one marriage giving rise to one branch and the other to the other branch (as in the half-cousins mentioned in section 3.4). According to table 3.2 these pedigrees should be rare and they will be treated like the standard ones.

The number of intermediate ancestors varies from one (in uncle–niece and aunt–nephew) to any number, but in practice we consider only consanguinities up to third cousins or equivalent, which have three intermediate ancestors in each of the two branches, for a total of six. In general, the number of intermediate ancestors is $i + j$.

Left and right branches have equal length in even cousins, so $i = j$. In uneven cousins it is important to distinguish if the husband is in the shorter (HSB = husband in shorter branch) or longer branch (WSB = wife in shorter branch).

The sex of intermediate ancestors has an important influence on the expected and observed frequencies of consanguineous marriages. We shall therefore distinguish consanguineous pedigrees on the basis of the sex of intermediate ancestors. For each degree there is a number of pedigrees equal to 2 to the power $(i + j)$, since there are $i + j$ intermediate ancestors who can be of either sex. But there are two types of uneven cousins pedigrees, HSB and WSB, so the total number pedigrees of 1½ cousins, HSB + WSB, for instance, is equal to the number of pedigrees of second cousins.

The $i$, $j$, and number of pedigrees $n_p$, for the six consanguinity degrees is, therefore,

|  | $i$ | $j$ | $n_p$ |
|---|---|---|---|
| AN, UN | 0 | 1 | 4 (2 HSB + 2 WSB) |
| First cousins | 1 | 1 | 4 |
| 1½ cousins | 1 | 2 | 16 (8 HSB + 8 WSB) |
| Second cousins | 2 | 2 | 16 |
| 2½ cousins | 2 | 3 | 64 (32 HSB + 32 WSB) |
| Third cousins | 3 | 3 | 64 |

The number $n_p$, of pedigrees is also equal to the number $n_c$ of consanguineous individuals participating to the next generation in a stationary population. In a population increasing or decreasing by a factor expressed by the net growth rate per generation $n_r$, the number $n_c$ of consanguineous individuals of degree $i + j$ is equal to

$$n_p \times n_r^{(i+j)/2}$$

In the following we will ignore this complication, which introduces a relatively small correction factor, compared with the majority of others.

Our interest of distinguishing all pedigrees is that it allows us to take into account the sex of intermediate ancestors, which is responsible for two effects of some magnitude: the effect of spouses' age and the sex differential in migration.

### Effect of Spouses' Age

Husbands tend to be older than wives. In a direct survey in Parma we found

average age of men at childbirth  $h = 33.2$ years
average age of women at childbirth  $w = 28.7$ years

and the difference $d = h - w = 33.2 - 28.7 = 4.5$ is approximately equal to the average difference of age at marriage of husband and wife.

On average, the age difference between two sibs giving rise to the two lineages is zero (since the order of birth of sibs is immaterial), and the difference expected between two prospective consanguineous mates depends on the sexes of the intermediate ancestors on the two lines. For instance, in the marriage between two second cousins, in which the intermediate ancestors of the husband are two females and those of the wife are also two females, the expected age difference between husband and wife is zero, because the average age at which the intermediate ancestors have had children is twice $w$ (two women generations) on the side of the husband and the same on the side of the wife. But if on the side of the wife there is one male and one female, and, as before, these are two females on the side of the husband, then the prospective wife is expected to be born $h + w$ years after the sibs, while the prospective husband is expected to be born $2w$ years after the sibs. The expected age difference between husband and wife in this pedigree is, therefore, $h + w - 2w = h - w = d$. This is exactly the ideal age difference between husband and wife. One can expect this pedigree to give rise to the highest number of marriages on the basis of age difference alone, among those that are otherwise comparable.

The same computation can be done for all pedigrees, and thus one can evaluate the expected age difference between husband and wife in each pedigree type. This can then be compared with the observed number of marriages to test the importance of the age difference factor in marriage. A more sophisticated calculation is shown in the next chapter. Here we show that the expected age difference, calculated by comparing the numbers of male and female generations in the branches leading to the husband and wife, has a clear-cut effect on the frequency of the pedigree type.

The greatest age effect is, of course, observed when comparing the greatest age differences. These are found for marriages between aunts and nephews, in which the husband branch has an expected birth date of $h$ or $w$, depending on whether the nephew is the child of a brother or sister of his aunt, and the aunt has zero ancestors. The expected age difference is $-h$ or $-w$, respectively, and one can expect this to be the least likely marriage, as is, indeed, observed (see table 3.10).

## Sex Differential in Migration

We have mentioned that more remote relatives are scattered over a wider area, with every generation since the common ancestors adding further distance (geographic or social) to the relatives, and hence decreasing their chance of marriage. Geographic distance between mother and offspring is greater than that between father and offspring (Cavalli-Sforza [1962] for the Parma region), because in an agricultural economy it is mostly males who inherit the land, and females move at marriage to the husband's residence. In the same way, especially among more remote consanguineous marriages, pedigree types with a greater number of intermediate female ancestors have a lesser chance of marriage.

In tables 3.5–3.10 we make a first quantitative test of these two hypotheses, which can be further refined later, by analyzing the frequencies of the various pedigree types within a consanguinity degree. One can classify all pedigrees according to the $h - w$ age difference, and also according to the sex of intermediate ancestors. Many ped-

TABLE 3.5. Parma and Piacenza (1851–1917)—third cousins: matrix of types of pedigrees

| | | | | | $K$ | | | | | Total | Mean frequency |
|---|---|---|---|---|---|---|---|---|---|---|---|
| Δ | | 0 | 1 | 2 | 3 | 4 | 5 | 6 | | | |
| −3d | | — | — | — | 21 : 26 | — | — | — | | 26 | 26.00 |
| −2d | | — | — | 5 : 16<br>17 : 14<br>20 : 16<br>**15.33** | — | 23 : 39<br>29 : 31<br>53 : 31<br>**33.67** | — | — | | 147 | 24.50 |
| −d | | — | 1 : 10<br>4 : 14<br>16 : 19 | — | 7 : 32  28 : 36<br>13 : 30  37 : 15<br>19 : 19  49 : 22<br>22 : 19  52 : 25<br>25 : 25 | — | 31 : 70<br>55 : 67<br>61 : 77 | — | | 480 | 32.00 |
| | | | **14.33** | | **24.78** | | **71.33** | | | | |

| Δ | | | | | | | | Total | Mean frequency |
|---|---|---|---|---|---|---|---|---|---|
| 0 | 0 : 21<br>**21** | — | 3 : 32  24 : 19<br>6 : 21  33 : 21<br>9 : 17  36 : 15<br>12 : 16  48 : 27<br>18 : 18<br>**20.67** | — | 15 : 70  51 : 39<br>27 : 40  54 : 25<br>30 : 49  57 : 37<br>39 : 37  60 : 53<br>45 : 44<br>**43.78** | — | 63 : 167<br>**167** | 768 | **38.40** |
| d | — | 2 : 17<br>8 : 20<br>32 : 24<br>**20.33** | — | 11 : 23  41 : 28<br>14 : 31  44 : 35<br>26 : 23  50 : 22<br>35 : 30  56 : 31<br>38 : 22<br>**27.22** | — | 47 : 77<br>59 : 67<br>62 : 91<br>**78.33** | — | 541 | **36.07** |
| 2d | — | — | 10 : 26<br>34 : 22<br>40 : 16<br>**21.33** | — | 43 : 43<br>46 : 40<br>58 : 41<br>**41.33** | — | — | 188 | **31.33** |
| 3d | — | — | — | 42 : 34 | — | — | — | 34 | **34.00** |
| Total | 21 | 104 | 296 | 528 | 619 | 449 | 167 | 2184 | **34.00** |
| Mean frequency | **21.00** | **17.33** | **19.73** | **26.40** | **41.27** | **74.83** | **167.00** | | **34.13** |

*Note.* Rows are categories of expected age difference, Δ; columns are male numbers among intermediate ancestors, *k*. Code of pedigree in italic, frequency in roman type. Numbers in bold are the average frequencies of each cell of the matrix.

TABLE 3.6. Parma and Piacenza (1851–1917)—second cousins once removed: matrix of types of pedigrees

| Δ | K | | | | | | Total | Mean frequency |
|---|---|---|---|---|---|---|---|---|
| | 0 | 1 | 2 | 3 | 4 | 5 | | |
| **Bride in the shorter branch** | | | | | | | | |
| −h − 2d | — | — | — | 42 : 2<br>**2** | — | — | 2 | **2.00** |
| −h − d | — | — | 10 : 3<br>34 : 2<br>40 : 4<br>**3.00** | — | 43 : 7<br>46 : 7<br>**7.00** | — | 23 | **4.60** |
| −h | — | 2 : 7<br>8 : 2<br>32 : 3<br>**4.00** | — | 11 : 14  38 : 4<br>14 : 11  41 : 10<br>35 : 15  44 : 15<br>**11.50** | — | 47 : 27<br>**27** | 108 | **10.80** |
| −h + d | 0 : 9<br>**9** | — | 3 : 12  12 : 12<br>6 : 2  33 : 11<br>9 : 11  36 : 13<br>**10.17** | — | 15 : 31<br>39 : 26<br>45 : 22<br>**26.33** | — | 149 | **14.90** |
| −h + 2d | — | 1 : 9<br>4 : 11<br>**10.00** | — | 7 : 26<br>13 : 25<br>37 : 38<br>**29.67** | — | — | 109 | **21.80** |
| −h + 3d | — | — | 5 : **29** | — | — | — | 29 | **29.00** |
| Total | 9 | 32 | 99 | 160 | 93 | 27 | 420 | 13.13 |
| Mean frequency | **9.00** | **6.40** | **9.90** | **16.00** | **18.60** | **27.00** | | |

**Groom in the shorter branch**

| | | | | | | | Total | Mean frequency |
|---|---|---|---|---|---|---|---|---|
| *h* − 3*d* | — | — | *21* : **52** | — | — | — | 52 | **52.00** |
| *h* − 2*d* | — | *17* : 25<br>*20* : 26<br>**25.50** | — | *23* : 51<br>*29* : 53<br>*53* : 57<br>**53.67** | — | — | 212 | **42.20** |
| *h* − *d* | *16* : 20<br>**20** | — | *19* : 28  *28* : 28<br>*22* : 28  *49* : 28<br>*25* : 32  *52* : 29<br>**28.83** | — | *31* : 78<br>*55* : 58<br>*61* : 87<br>**74.33** | — | 416 | **41.60** |
| *h* | — | *18* : 11<br>*24* : 26<br>*48* : 18<br>**18.33** | — | *27* : 28  *54* : 31<br>*30* : 29  *57* : 32<br>*51* : 25  *60* : 33<br>**29.67** | — | *63* : 100<br>**100** | 333 | **33.00** |
| *h* + *d* | — | — | *26* : 5<br>*50* : 9<br>*56* : 19<br>**11.00** | — | *59* : 37<br>*62* : 38<br>**37.50** | — | 108 | **21.60** |
| *h* + 2*d* | — | — | — | *58* : 6<br>**6** | — | — | 6 | **6.00** |
| Total | 20 | 106 | 258 | 345 | 298 | 100 | 1127 | |
| Mean frequency | **20.00** | **21.20** | **25.80** | **34.50** | **59.60** | **100.00** | | **35.22** |

*Note.* Rows are categories of expected age difference, Δ; columns are male numbers among intermediate ancestors, *k*. Code of pedigree in italic, frequency in roman type. Numbers in bold are the average frequencies of each cell of the matrix.

TABLE 3.7. Parma and Piacenza (1851–1957)—second cousins:
matrix of types of pedigrees

| Δ | K 0 | 1 | 2 | 3 | 4 | Total | Mean frequency |
|---|---|---|---|---|---|---|---|
| −2d | — | — | 5 : 237 | — | — | 237 | **237.00** |
| −d | — | 1 : 198<br>4 : 200<br>**199.0** | — | 7 : 381<br>13 : 368<br>**374.5** | — | 1147 | **286.75** |
| 0 | 0 : 231<br><br>**231** | — | 3 : 344<br>6 : 268<br>9 : 267<br>12 : 289<br>**292.0** | — | 15 : 674<br><br>**674** | 2073 | **345.50** |
| d | — | 2 : 231<br>8 : 220<br>**225.5** | — | 11 : 437<br>14 : 383<br>**410.0** | — | 1271 | **317.75** |
| 2d | — | — | 10 : 260 | — | — | 260 | **260.00** |
| Total | 231 | 849 | 1665 | 1569 | 674 | 4988 | |
| Mean frequency | **231.00** | **212.25** | **277.50** | **392.25** | **674.00** | | **311.75** |

*Note.* Rows are categories of expected age difference, Δ; columns are male numbers among intermediate ancestors, *k*. Code of pedigree in italic, frequency in roman type. Numbers in bold are the average frequencies of each cell of the matrix.

igree types within the same degree give the same expectations for one, the other, or both of these classifications, thus simplifying the analysis. Even though it is superficially the most complex method, it pays to start the analysis with third cousins, which are given for the dioceses of Parma and Piacenza combined, 1851–1917 in table 3.5.

The two criteria of classification of the 64 pedigrees, numbered from 0 to 63 (whose pedigrees and codes are given in full in figure 3.3 b), are as follows:

1. On the basis of age differences, one can generate seven categories, forming the seven rows of table 3.5b. The expected age difference Δ between husband and wife is given in the first column and is calculated according to formula 3.2. The minimum $\Delta = -3d$, (first row) is found in only one pedigree, code 21, of which there were 26 examples in the total number of 2184 observed pedigrees. The maximum $\Delta = 3d$ is observed in pedigree code 42, with 34 observed

TABLE 3.8. Parma and Piacenza (1851–1957)—First cousins once removed: matrix of types of pedigrees

| Δ | K | | | | Total | Mean frequency |
|---|---|---|---|---|---|---|
| | 0 | 1 | 2 | 3 | | |
| **Bride in the shorter branch** | | | | | | |
| $-h - d$ | — | — | $10 : 18$ | — | 18 | **18.00** |
| $-h$ | — | $2 : 12$ $8 : 27$ **19.5** | — | $11 : 49$ **49** | 88 | **29.33** |
| $-h + d$ | $0 : 31$ **31** | — | $3 : 56$ $9 : 55$ **55.5** | — | 142 | **47.33** |
| $-h + 2d$ | — | $1 : 80$ | — | | 80 | **80.00** |
| Total | 31 | 119 | 129 | 49 | 328 | |
| Mean frequency | **31.00** | **39.67** | **43.00** | **49.00** | | **41.00** |
| **Groom in the shorter branch** | | | | | | |
| $h - 2d$ | — | $5 : 231$ | — | — | 231 | **231.00** |
| $h - d$ | $4 : 130$ **130** | — | $7 : 207$ $13 : 175$ **191.0** | — | 512 | **170.67** |
| $h$ | — | $6 : 90$ $12 : 90$ **90.0** | — | $15 : 183$ **183** | 363 | **121.00** |
| $h + d$ | — | — | $14 : 60$ | — | 60 | **60.00** |
| Total | 130 | 411 | 442 | 183 | 1166 | |
| Mean frequency | **130.00** | **137.00** | **147.33** | **183.00** | | **145.75** |

*Note.* Rows are categories of expected age difference, Δ; columns are male numbers among intermediate ancestors, $k$. Code of pedigree in italic, frequency in roman type. Numbers in bold are the average frequencies of each cell of the matrix.

TABLE 3.9. Parma and Piacenza (1851–1957)—first cousins:
matrix of types of pedigrees

| | | K | | | Mean |
| Δ | 0 | 1 | 2 | Total | frequency |
|---|---|---|---|---|---|
| −d | — | *1* : **828** | — | 828 | **828.00** |
| *0* | *0* : **1,087** | — | *3* : **900** | 1,987 | **993.50** |
| *d* | — | *2* : **1,149** | — | 1,149 | **1,149.00** |
| Total | 1,087 | 1,977 | 900 | 3,964 | |
| Mean frequency | **1,087.00** | **988.50** | **900.00** | | **991.00** |

Note. Rows are categories of expected age difference, Δ; columns are male numbers among intermediate ancestors, $k$. Code of pedigree in italic, frequency in roman type. Numbers in bold are the average frequencies of each cell of the matrix.

pedigrees. The last column shows the average frequency of the pedigrees with Δ varying from −3$d$ to 3$d$. The maximum frequency is observed for Δ = 0, with a mean value of 38.4 pedigrees. This value is not much greater than that of the pedigrees generating the most satisfactory age difference $d$ between mates. We will try to explain this discrepancy later.

TABLE 3.10. Parma and Piacenza (1851–1957)—uncle–niece and aunt–nephew: matrix of types of pedigrees

| | | K | | | Mean |
| Δ | 0 | 1 | 2 | Total | frequency |
|---|---|---|---|---|---|
| Aunt–nephew | | | | | |
| −$h$ | — | *2* : **1** | — | 1 | **1.00** |
| −$h + d$ (−$w$) | *0* : **9** | — | — | 9 | **9.00** |
| Uncle–niece | | | | | |
| $h − d$ ($w$) | — | *1* : **64** | — | 64 | **64.00** |
| $h$ | — | — | *3* : **31** | 31 | **31.00** |
| Total | 9 | 65 | 31 | 105 | |
| Mean frequency | **9.00** | **32.50** | **31.00** | | **26.25** |

Note. Rows are categories of expected age difference, Δ; columns are male numbers among intermediate ancestors, $k$. Code of pedigree in italic, frequency in roman type. Numbers in bold are the average frequencies of each cell of the matrix.

2. On the basis of the number of males ($k = m_i + m_j$) among the six intermediate ancestors one can generate seven categories ($k = 0$, $1, \ldots, 6$), forming the seven columns. Numbers of pedigrees are smallest with fewer males, because all or almost all intermediate ancestors are females and they have spread widely enough that fewer marriages were possible (21 pedigrees for $k = 0$). The observed frequencies increase with increasing number of males and reach a maximum for $k = 6$ (167 pedigrees).

Tables 3.6–3.10 show the other consanguinity degrees. In uneven cousins it is convenient to express the expected age difference in terms of two parameters, $d$ and $h$. The age effect is always clear: dramatic in the comparison of AN and UN and in other uneven cousins, where it is expected to be greater than in even cousins.

The effect of the sex of intermediate ancestors is clearly visible in second and third cousins, but also in 1½ and 2½ cousins: an increase of the number of males is accompanied by an increase in frequency for all degrees, as expected because of the greater dispersion of female ancestors. But in the very short pedigrees, UN, AN, and first cousins, the effect of sex is reversed. Here there is serious confounding of the two effects, age and migration, and probably also of another social effect, discussed in Barrai et al. (1962), which is that female ancestors tend to keep closer contacts among relatives and favor marriages among them, thus partially eliminating the effect of migration. This phenomenon is explained in section 4.3. In summary, the most remote consanguinity degrees are more powerful in showing the effects being studied: in uneven cousins the age effect and in even cousins the effect of differential migration of the two sexes, as revealed by the sex of intermediate ancestors.

Pedigrees belonging to the same class of age difference and sex of intermediate ancestors tend to be homogeneous in frequency. Heterogeneity chi squares among pedigree types with the same $\Delta$ and $k$ (not shown in the tables, but given in Barrai et al. [1962]) are not significant, except for two cases: pedigree 15 in third cousins for $\Delta = 0$ and $k = 4$ is significantly more frequent than the other pedigrees of the group; and the same is true for pedigree 3 in second cousins for $\Delta = 0$ and $k = 2$. In the next chapter this discrepancy is explained as an effect of the role of female ancestors in the pedigree. By contrast, the effects of age difference and of the sex of intermediate ancestors are almost always highly significant in all tables. In the following chapter, the theories of age effects and migration are further characterized and tested against the same data and are used in other circumstances to explain observations and understand exceptions.

Tables 3.5–3.10 were built with a convention slightly different from that used by Barrai et al. (1962). Here and there the $\Delta$ value is an expected age difference between mates in the pedigree indicated in any given row, but the origin is set differently. Here the $\Delta$ value is taken equal to the expected age difference between mates, which depends on the number of male and female ancestors in the two branches leading to the two consanguineous spouses. The formula used is:

$$\Delta = (m_j h + f_j w) - (m_i h + f_i w) \tag{3.2}$$

In Barrai et al. (1962) the convention used was to set the origin of the $\Delta$ scale where the expected value is equal to a difference of $d$ between the male and the female spouses. This is the value at which we may expect the number of pedigrees to be a maximum because the age difference is that viewed as optimal for marriage. In fact, for statistical reasons, but also because of a bias that we have recently discovered (explained in chapter 4), this is not always the case. Moreover, the sign of the $\Delta$ scale is inverted with respect to our use.

# Probability of
# Consanguineous Marriages

## 4.1 THEORY OF AGE EFFECTS ON THE FREQUENCY
## OF CONSANGUINEOUS MARRIAGES

In general, two consanguineous individuals of a given pedigree are expected to have a distribution of difference between their ages that depends on the degree of consanguinity and type of pedigree. Some have, on average, an extreme age difference, greater than that which is most common in the general population (and in some cases in opposite direction, as, for example, for aunt and nephew). Pedigrees of even cousins have an expected age difference near zero, and therefore closer to the usual one. In standard, nonconsanguineous marriages, the age of spouses shows a high correlation, the husband being on average older by a small number of years. Marriages in which the age difference is far from that norm are rarer, and the more so, the greater the age difference from the norm. But there is considerable variation in the actual age difference between spouses in individual marriages.

The analysis of age effects on consanguineous marriages was carried out by Barrai et al. (1962) on consanguineous marriages of the North Emilia bishoprics. It included a joint analysis of both age effects and migration. A first attempt at a more comprehensive mathematical treatment, using the approximation made possible by normal distributions of ages at marriage, was the subject of a thesis at the University of Parma by Braglia (1962). This attempt is very similar to the slightly more complete treatment of age given by Hajnal (1963). The most complete treatment including age, migration, and other fac-

TABLE 4.1. Demographic quantities necessary for age effects analysis and estimates obtained from a sample of the diocese of Parma

| | | |
|---|---|---|
| $h$ | mean generation time (mean age at childbirth) of fathers | = 33.02 years |
| $w$ | mean generation time (mean age at childbirth) of mothers | = 28.81 |
| $d$ | mean age difference between husband and wife | = 4.21 |
| $V_m$ | individual variance of $h$, the male age at child birth | = 62.61 |
| $V_f$ | individual variance of $w$, the female age at childbirth | = 38.61 |
| $V_d$ | variance of individual values of $d$ | = 24.00 |
| $V_s$ | variance of difference of age between sibs | = 25.78; 50.61 |

tors appeared in 1966 (Cavalli-Sforza et al.). Further contributions were the prediction of the bias in age at marriage and comparison with real data (Zei et al. 1971). The demographic analysis of age effects requires knowledge of seven quantities, which are listed in table 4.1, together with estimates obtained from a sample of the living populations of Parma, examined in 1958 (Mainardi et al. 1962).

The variance of the age difference between sibs ($V_s$) is given in table 4.1 with two values. The first value (25.78) was obtained from the ages of all sibs born in large families of the living Parma population and was used in earlier work. Later, we developed doubts on its validity, since the sibs who marry and settle in the area are a subsample of all those born, which may increase the value of $V_s$. Attempts at estimating it from data of frequencies of consanguineous marriages strengthened this suspicion, and led us to chose as the value of $V_s$ the average of $V_m$ and $V_f$ which is almost twice the first value ($V_s$ = 50.61, second value given in table 4.1). The estimate of $V_s$ remains the weakest among those above.

We can now repeat in a more general form the calculation of the expected age difference between consanguineous husband and wife given in section 3.4. The expected birth date of the wife (at the right in the pedigree for even cousins, in the $j$ branch) is equal to $m_j$ male generations + $f_j$ female generations, there being $m_j$ male and $f_j$ female ancestors in the lineage giving rise to the wife. The birth date of the wife is thus $m_j h + f_j w$ years, and that of the husband at the left (in the $i$ branch) is $m_i h + f_i w$ years. The expected age difference between the two sibs (i.e., the children of the common ancestor) is zero with a variance $V_s$. The wife is therefore younger than the husband by the quantity already given in 3.2:

$$(m_j\, h + f_j\, w) - (m_i\, h + f_i\, w)$$

so that the expected (or mean) age difference between the consanguineous husband and wife can also be written

$$M_{\text{diff}} = h(m_j - m_i) + w(f_j - f_i) \tag{4.1}$$

and its variance will be

$$V_{\text{diff}} = V_s + (m_i + m_j)V_m + (f_i + f_j)V_f \tag{4.2}$$

We assume that the distribution of age differences among consanguineous individuals of given type, $F(x)$, is normal, with the mean given by (4.1) and variance by (4.2). Even if this may not be necessarily true for close consanguinities, for more remote ones the distribution will tend to become closer and closer to normal because of the central limit theorem.

We want to use these formulas for two kinds of prediction. The first is that *the age of potential consanguineous mates may be quite different from the ideal one at marriage.* The extreme case we have seen is that of UN and AN. Nevertheless, some of these marriages take place. Therefore, the age of consanguineous spouses may differ from that observed in the general population.

Marriages among consanguineous individuals whose age difference is less remote from that favored by the general population will be more likely to take place, but, even so, there will be a considerable distortion of the average age at marriage in consanguineous marriages. One can predict the bias thus arising if one assumes that consanguineous mates try to follow the general trend in the attraction for age difference of spouses; that is, if one assumes that the same probability of choice based on age difference applies in consanguineous as well as in general marriages.

Zei et al. (1971) made this calculation by assuming that the distribution $G(x)$ of the age difference of husband and wife in the general population is normal with mean equal to $d$ and variance $V_d$. We know this is not exact, but this approximation is not too serious and greatly simplifies the computations.

We know that $F(x)$, the distribution of age difference between two consanguineous individuals, is very nearly normal with the mean and variance given earlier. On the assumption of normality of both $G(x)$ and $F(x)$, the distribution of the expected age difference $\Delta$ of consanguineous spouses is the product of two normal distributions $B(x) = F(x) G(x)$, which is also normal and has easily calculated mean and variance. The mean of $B(x)$ is

$$\Delta_{exp} = \frac{M_{diff} V_d + d V_{diff}}{V_d + V_{diff}} \tag{4.3}$$

Calculation of the expected age differences between husband and wife for a number of consanguineous marriages and their comparison with the observed age differences are given in table 4.2. Data are from Vatican archives and come from all Italy, 1911–1964 (see chapter 9).

The differences between expected and observed $\Delta$ between the ages of consanguineous husband and wife are very close to zero, with the exception of uncle–niece and aunt–nephew. Nevertheless, these differences are *significantly* different from zero, due to the high number of marriages. The test applied is:

$$t = \frac{\Delta_{obs} - \Delta_{expt}}{\sqrt{V_d/N}}$$

where $N$ is the number of marriages and the variance is that obtained from data of the Parma diocese, $V_d = 24$ (see table 4.1). We know that this variance varies greatly according to regions and this contributes to generating a significant discrepancy between observed and expected age difference. The aunt–nephew case may be especially sensitive to the approximation of using a normal distribution of the difference in age at marriage, or it may involve greater social difficulties than other consanguineous marriages, especially because of the extreme age difference. This can help explain why so few aunt–nephew marriages occur. Only rarely is an aunt younger than her nephew, and only a few nephews accept an older aunt as a wife.

The second problem in which we are interested is *to calculate the effect on the number of expected marriages due to age differences between relatives*, and compare it with the observed number of marriages. What is needed is the double integral:

$$J = \int\int F_m(t, t') \, F_c(t, t') \, dt \, dt' \tag{4.4}$$

where $F_m(t, t')$ is the joint distribution of ages at marriage of males and females $t, t'$, and $F_c(t, t')$ that of relatives of ages $t$ and $t'$. Again the normal approximations will be used. Then the integral [4.4] reduces to

$$J = \frac{\exp(-M^2/2S^2)}{\sqrt{2\pi S^2}} \tag{4.5}$$

where

$$M = M_{\text{diff}} - d$$
$$S^2 = V_{\text{diff}} + V_d$$

This result (Cavalli-Sforza et al. 1966) is similar to one published by Hajnal (1963).

The normal approximation introduces an error because the distribution of ages at marriage are not exactly normal and the correlation is not strictly linear. The error thus introduced in the $J$ values was evaluated using a better approximation (a log-normal distribution of the ages at marriage) and was found to be at most 14% for uncle–niece and aunt–nephew marriages, but less than 2% for first-cousin marriages (Cavalli-Sforza et al. 1966). Then, the error is large only for the most asymmetric marriages, which are rare; for the other degrees the improvement obtained with the better approximation is not so striking as to justify the complication introduced in the calculations.

These formulas have been used in a number of computations of expected frequencies of pedigree types (see, e.g., Cavalli-Sforza et al. 1966) and, in general, give good results. It is necessary, however, to introduce the expectations under migration, which are discussed in the next section.

TABLE 4.2. Prediction of the bias in age at consanguineous marriage: expected and observed mean difference of age in years between husband and wife in various types of consanguineous marriages (data from all Italy)

| Marriage | Type of pedigree | $i$ | $j$ | $m_i$ | $m_j$ | Age difference of mates | | | |
| --- | --- | --- | --- | --- | --- | --- | --- | --- | --- |
| | | | | | | Expected (years) | Observed (years) | No. of marriages | Difference (expt. − obs.) |
| Uncle–niece | 1 | 0 | 1 | 0 | 0 | 10.89 | 11.33 | 2,877 | −0.44** |
| | 3 | 0 | 1 | 0 | 1 | 10.36 | 12.23 | 2,071 | −1.87** |
| Aunt–nephew | 0 | 1 | 0 | 0 | 0 | −4.76 | 1.65 | 399 | −6.41** |
| | 2 | 1 | 0 | 1 | 0 | −3.74 | −0.54 | 86 | −3.20** |
| First cousins | 0 | 1 | 1 | 0 | 0 | 3.41 | 3.50 | 11,876 | −0.09 |
| | 2 | 1 | 1 | 0 | 1 | 4.21 | 3.98 | 9,602 | +0.23** |
| | 1 | 1 | 1 | 1 | 0 | 2.87 | 3.08 | 7,753 | −0.21** |
| | 3 | 1 | 1 | 1 | 1 | 3.63 | 3.76 | 37,404 | −0.13* |

| | | | | | | | | |
|---|---|---|---|---|---|---|---|---|
| First cousins once removed (husband in shorter branch) | 4 | 1 | 2 | 0 | 0 | 7.78 | | |
| | 12, 6 | 1 | 2 | 0 | 1 | 7.86 | | |
| | 14 | 1 | 2 | 0 | 2 | 7.92 | | |
| | 5 | 1 | 2 | 1 | 0 | 6.79 | 7.41 | |
| | 13, 7 | 1 | 2 | 1 | 1 | 6.97 | | |
| | 15 | 1 | 1 | 1 | 2 | 7.12 | | |
| | | | | | | | 6.20 | 6,733 +1.21** |
| First cousins once removed (wife in shorter branch) | 0 | 2 | 1 | 0 | 0 | −0.58 | | |
| | 1 | 2 | 1 | 0 | 1 | 0.56 | | |
| | 8, 2 | 2 | 1 | 1 | 0 | −0.50 | 0.00 | |
| | 9, 3 | 2 | 1 | 1 | 1 | 0.50 | | |
| | 10 | 2 | 1 | 2 | 0 | −0.45 | | |
| | 11 | 2 | 1 | 2 | 1 | 0.45 | | |
| | | | | | | | 0.98 | 1,438 −0.98** |

Source. Zei et al. (1971).
*Significantly different from zero at $P = 0.01$.
**Significantly different from zero at $P = 0.001$.

## 4.2 MIGRATION AS A FACTOR AFFECTING THE
## FREQUENCY OF CONSANGUINEOUS MARRIAGES

It is clear from the preliminary analysis in section 4.1 that age at marriage plays an important role in explaining the difference in frequency of consanguineous marriages, but is not sufficient to explain it all. It was shown in chapter 3 that an additional important factor is migration. In particular, differential migration of the two sexes can explain a substantial fraction of the residual variation after age effects are taken into account. Marriages tend to occur among individuals who reside and, in general, are born at a short distance from one another. There exist many studies on the distribution of geographic distances among birthplaces of spouses. Many of them are based on data from parish books and are therefore relevant for the period or periods here considered (see Cavalli-Sforza [1962], Bodmer and Cavalli-Sforza [1968], and Wijsman and Cavalli-Sforza [1984] for summaries). Especially in agricultural economies, marriage is a major cause of migration and is predominantly "patrilocal" where, as is most often the case, males inherit the land. In other words, it is mostly women who move to the husband's residence when they marry. It is usually only in families with no male children that a woman may inherit the land and her husband changes residence at marriage. In Barrai et al. (1962), we examined quantitatively the effects of migration on the simple hypothesis that there is a constant ratio between the probability of migration of females and males. If $p_m$ is the probability that males do NOT emigrate, and $p_f$ the same for females, a consanguineous marriage will take place only if the intermediate ancestors of the spouses have not emigrated from the area of origin. This probability is simply given by

$$p_m{}^m \times p_f{}^f \qquad\qquad (4.6)$$

where $m = m_i + m_j$ and $f = f_i + f_j$ are the total males and females, respectively, among the intermediate ancestors. Knowing the ratio $p_m/p_f = c$, we can compare the probability of pedigrees within the same consanguinity degree. The approach allows only an approximate fit of the observations; migration was accurately represented for a high number of males, and less so for a high number of females.

More accurate models were put forward. Unfortunately, no simple distribution fitting the data lends itself to the necessary mathematical manipulations for taking account of the distribution of migration at marriage over many generations. Two simple models tested, a sum of normal bivariate distributions and a sum of exponentials, gave reasonable fits to migration distributions for one generation. The first model could be extended analytically to a multiple-generations approach, necessary for dealing with the accumulation of the migration events of many ancestors over several generations. The second had to be analyzed by numerical integration. Both showed that the simpler probability model, similar to the one used in the 1962 paper, could reasonably approximate the expectations of the more elaborate multivariate normal and exponential distributions. According to this model, two individuals mate with probability given by (4.6) if migration during their lifetime keeps them within a specific range, which can be considered as a "mating range."

Estimates of $p_m$ and $p_f$ are available approximately from distributions of geographic distance among birthplaces of mates. An estimate of $p_m$ and $p_f$ from such curves gives values of $c = p_m/p_f = 1.13$ (Cavalli-Sforza et al. 1966). Direct estimates of $c$ from pedigrees of third cousins are as high as $c = 1.5$, those of second cousins are closer to 1.3, while in first cousins the situation seems to be reversed ($c = 0.95$) because other social factors, discussed later, prevail.

If $p_m/p_f$ were constant the probabilities $P$ of consanguineous marriages for the various degrees would be expected to rise almost exponentially with the number of males among intermediate ancestors and the curves of log of $P$ would be straight and parallel (as shown in figure 3 of Cavalli-Sforza et al. [1966]). The change in slope, negative with first cousins and positive in second and third cousins, for the observed data shown in figure 4.1 is probably the consequence of a phenomenon discussed in section 4.3, which may be the main cause of the deviation from the expected linearity and parallelism. It is also likely that $p_m$ and $p_f$ changed over time and space, and the lack of linearity may be due to it. Not surprisingly, the effect of the number of intermediate ancestors is less dramatic in odd cousins, taking account of the difference between pedigrees in which male or female are in the shorter branch (figures 4.2 and 4.3).

In conclusion, the sex of intermediate ancestors has an important effect in determining the occurrence of consanguineous marriages, due, at least in part, to the lesser dispersal of males over generations,

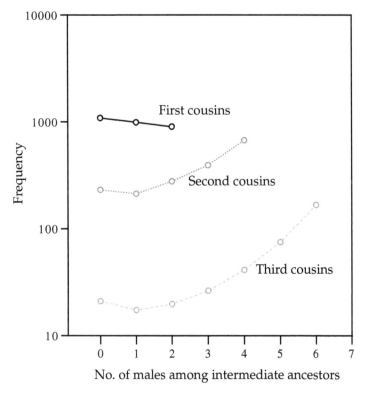

FIGURE 4.1. Frequency of pedigrees as a function of the number of males among intermediate ancestors in first, second, and third cousins.

compared with females who disperse more. However, this is not sufficient to fully explain results quantitatively, and the effect of the sex of intermediate ancestors tends to disappear for close consanguinities or when intermediate ancestors are mostly female. We see in the next section that other social factors must play a role.

## 4.3 THE ROLE OF WOMEN IN MAINTAINING FAMILY TIES AMONG RELATIVES

Another analysis helped to throw some light on the problem of the nonlinearity of the curves of figure 4.1, and we summarize it and extend it from Barrai et al. (1962). It shows the very likely presence

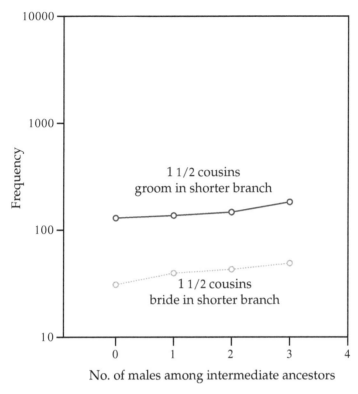

FIGURE 4.2. Frequency of pedigrees as a function of the number of males among intermediate ancestors in first cousins once removed (1½).

of another factor: the maintenance of family ties by women relatives. More advanced study of this and other related factors would require a sociological analysis, and we hope our findings will stimulate sociologists to investigate it further.

Two possible causes of the departure from the simplest expectations regarding the effect of the sex of intermediate ancestors were considered: the sex of the remote ancestors (the sibs) and the sex of the nearest ancestors, for example, the parents of the consanguineous spouses. One possible reason for remote ancestors to have an effect in rural areas, as those investigated are predominantly, is that only one male child may inherit the land and therefore marry and have progeny. Others may have to emigrate. Hence, pedigrees in which the

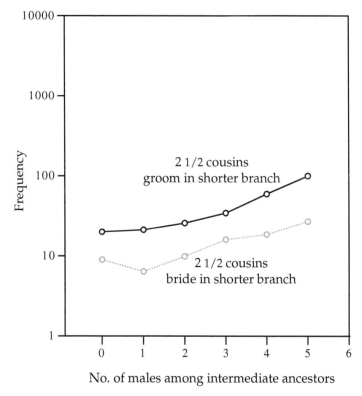

FIGURE 4.3. Frequency of pedigrees as a function of the number of males among intermediate ancestors in second cousins once removed (2½).

two earliest intermediate ancestors, the two sibs, are brothers may be more rare.

This hypothesis was tested on pedigrees of third cousins who had the same number of male intermediate ancestors and the same expected age difference. There are nine such pedigrees, all of which have four males out of six intermediate ancestors, and the same expected age difference ($\Delta = 0$). Observed numbers, grouped in 3 classes, are shown in table 4.3. With these numbers there is no significant tendency for pedigrees having two brothers as the oldest ancestors to be underrepresented.

The same pedigrees lend themselves, after regrouping (table 4.4), to the analysis of the other hypothesis: whether the sex of the parents

TABLE 4.3. Effect of the sex of the remotest intermediate ancestors on the frequency of third-cousin marriages

| Remotest ancestors (the sibs) | Pedigree codes | Expected proportion | Total expected numbers | Total observed numbers |
|---|---|---|---|---|
| Two brothers | 15, 27, 39, 51 | 4/9 | 175.1 | 186 |
| Brother–sister | 30, 45, 54, 57 | 4/9 | 175.1 | 155 |
| Two sisters | 60 | 1/9 | 43.8 | 53 |

$$\chi^2_{[2]} = 4.92$$
$$P = 0.08$$

of the consanguineous spouses has an effect. Here, the chi-square test with 2 df is 18.40, significant at $P = 0.1\%$. The most frequent pedigree (No. 15) is that in which the parents of the consanguineous spouses are two females and it is this class that gives the greatest contribute to the highly significant chi square.

We will not report here the details of the same study on second cousins. Here the two hypotheses cannot be separated: pedigrees in which the two oldest ancestors are brothers are the same as those in which the nearest ones are sisters. They are significantly the most frequent (344 versus 289 for marriages in which the parents of the spouses are both sisters and both brothers, respectively). The same study has even more restrictions if done for first cousins, where the nearest and remotest ancestors are the same, but again there is an advantage of marriages where parents of the consanguineous spouses are two sisters (1087 versus 900 with both brothers).

TABLE 4.4. Effect of the sex of the parents of consanguineous mates (the nearest intermediate ancestors) on the frequency of third-cousin marriages

| Nearest ancestors (the parents) | Pedigree codes | Expected proportion | Total expected numbers | Total observed numbers |
|---|---|---|---|---|
| Two brothers | 51, 54, 57, 60 | 4/9 | 175.1 | 154 |
| Brother–sister | 27, 30, 39, 45 | 4/9 | 175.1 | 170 |
| Two sisters | 15 | 1/9 | 43.8 | 70 |

$$\chi^2_{[2]} = 18.40,$$
$$P < 0.001$$

It seems reasonable to infer that when the parents of the spouses are females there is a greater probability that family contacts are maintained, thus increasing the probability of marriage. It is possible, in addition, that the mothers of the future spouses are instrumental in arranging the marriage, or at least in facilitating it. Since the third-cousins curve in figure 4.1 shows a slight deviation from linearity not only for 0 males among intermediate ancestors, but also for 1 male and perhaps for 2, it is possible that female grandparents also play a role in keeping family ties. Female intermediate ancestors are likely to compensate for their greater geographic distance from relatives, which inevitably rises with increasing remoteness of consanguinity, by avoiding disruption of social ties as often occur among relatives for financial or other disagreements.

Deviations from the theory concerning the effect of the number of male intermediate ancestors are also observed in the classes with the maximum number of males. Pedigree 63 with 6 males in third cousins and pedigree 3 with 4 males in second cousins show frequencies higher than those expected on the basis of linear increase (see figure 4.1). Failing other explications, the hypothesis of an incorrect classification of some pedigrees by the priests could be envisaged. Two consanguineous mates with the same surname could be erroneously considered as having only males connecting them with the common ancestor, therefore increasing the observed frequency of this class and, consequently, that of a difference of mates age, $\Delta = 0$, and causing a frequency loss for the other classes of pedigrees with $\Delta \neq 0$.

A further social factor that may influence the frequency of consanguineous matings is assortative mating for genetically or culturally inheritable traits (for instance, socioeconomic conditions). A theoretical study in Cavalli-Sforza et al. (1966) calculated the expectations among relatives, assuming that $r_c$ is the correlation between relatives for a heritable trait, for which there is assortative mating leading to a correlation $r_a$ between husband and wife. It was shown that the effect of the probability of consanguineous marriage due to assortative mating for heritable traits can be calculated as the factor

$$I_s = \frac{1}{1 - r_c \times r_a} \tag{4.7}$$

which can be greater or smaller than 1 depending on whether assortment is positive or negative. There are no estimates of $r_a$ or $r_c$ for these populations, but it seems likely that this factor is of no major significance, as it involves the product of two correlation coefficients.

## 4.4 OBSERVED AND EXPECTED FREQUENCIES OF MAJOR CONSANGUINITY DEGREES

The complexity of pedigree types makes it difficult to evaluate expectations for consanguinity degrees other than by computer. At the cost of a small approximation, however, it is possible to evaluate average expectations for all pedigree types within a specific consanguinity degree and to calculate formulas that allow a faster computation of correction factors and are simple enough to make their magnitude easier to evaluate.

The two major quantifiable factors determining expectations are age and migration. We calculate a factor determining expectation due to age, $f_{age}$, and a factor due to migration, $f_{mig}$, from parameters estimated for each consanguinity degree. These values were obtained for each pedigree and averaged for all pedigree types found in that degree. From $f_{age}$, $f_{mig}$, and the number $n_c$ of consanguineous individuals of that degree, which in a stationary population is equal to the number $n_p$ of pedigrees, we calculate a quantity proportional to the relative frequency of that consanguinity degree, $f_c(b)$, where $b$ is the degree itself. Since the first degree evaluated is the sum of AN + UN, and, in general, uneven cousins have two moieties, HSB and WSB, the two moieties are given the same $b$ value, but are for most purposes kept distinct, because they have different expectations.

In general,

$$f_c(b) = K \times n_c \times f_{age} \times f_{mig} \qquad (4.8)$$

$K$ is a proportionality factor, which for the purposes of this section is 1/sum of expectations for all degrees. $f_{age}$ is (see section 4.1)

$$f_{age} = \frac{\exp(-M^2/2S^2)}{\sqrt{2\pi S^2}} \qquad (4.9)$$

where

$$M = h(m_j - m_i) + w(f_j - f_i) - d$$

$$S^2 = V_s + (m_i + m_j)V_m + (f_i + f_j)V_f + V_d$$

$f_{mig}$ is (see section 4.2)

$$f_{mig} = p_m^b \times p_f^b = (p_m \times p_f)^b = P^b \qquad (4.10)$$

where $P = p_m \times p_f$. The values of $b$, $n_c$, $m_i$, $m_j$, $f_i$, $f_j$ for all degrees to be used in the calculations are given in table 4.5.

The use of formulas 4.8–4.10 involves a numerical approximation. In table 4.6 the exact values, calculated by summing the probability of all pedigree types for a given degree, are given along with the approximate ones, calculated with formulas [4.8–4.10]. The shorter method is especially useful for recalculating expectations for populations having other values of the demographic constants. Those used for table 4.6 are shown in table 4.1.

The observed frequencies obtained for the Parma diocese, limited to the period 1851–1917 (the only one for which data of all the degrees of interest are complete), are shown in the first columns of table 4.7, which is dedicated to fitting the Parma data with available theories. An attempt has been made to representing this set of observations, the best in terms of data quality (accuracy and completeness), on the basis of both best-known factors: age and migration. The age effects are calculated on the best set of background data: all

TABLE 4.5. Values of parameters to be used for all consanguinity degrees in approximate average formulas 4.8–4.10

| Degree | $b$ | $n_c$ | $m_i$ | $m_j$ | $f_i$ | $f_j$ |
|---|---|---|---|---|---|---|
| 1. Uncle–niece | 1 | 2 | 0 | 0.5 | 0 | 0.5 |
| 2. Aunt–nephew | 1 | 2 | 0.5 | 0 | 0.5 | 0 |
| 3. First cousins | 2 | 4 | 0.5 | 0.5 | 0.5 | 0.5 |
| 4. 1½ HSB | 3 | 8 | 0.5 | 1 | 0.5 | 1 |
| 5. 1½ WSB | 3 | 8 | 1 | 0.5 | 1 | 0.5 |
| 6. Second cousins | 4 | 16 | 1 | 1 | 1 | 1 |
| 7. 2½ HSB | 5 | 32 | 1 | 1.5 | 1 | 1.5 |
| 8. 2½ WSB | 5 | 32 | 1.5 | 1 | 1.5 | 1 |
| 9. Third cousins | 6 | 64 | 1.5 | 1.5 | 1.5 | 1.5 |

Note. HSB, husband in the shorter branch; WSB, wife in the shorter branch.

TABLE 4.6. Comparison of expectations for the product of $n_c \times f_{age} \times f_{mig}$, taking $P = 0.5$ for migration, calculated from the sum of expectations for all pedigrees, and from the approximation available in formula 4.8

| Degree | Approximate | Exact |
|---|---|---|
| 1. Uncle–niece | 0.022423 | 0.023188 |
| 2. Aunt–nephew | 0.002805 | 0.002852 |
| 3. First cousins | 0.310310 | 0.306543 |
| 4. 1½ HSB | 0.059548 | 0.063280 |
| 5. 1½ WSB | 0.018863 | 0.021090 |
| 6. Second cousins | 0.251803 | 0.246924 |
| 7. 2½ HSB | 0.080518 | 0.082998 |
| 8. 2½ WSB | 0.036381 | 0.039732 |
| 9. Third cousins | 0.217348 | 0.212394 |

*Note.* HSB, husband in the shorter branch; WSB, wife in the shorter branch.

demographic variables that are reasonably well known and do not change much in time and space.

The migration data are less solid, because the size of the "marriage range" is somewhat arbitrary and the estimates of migration we have made depend on it; $p_m$ and $p_f$ are the probabilities that males and females, respectively, do not migrate outside the marriage range in one generation. The distribution of geographic distance between father and offspring, or mother and offspring, in the Parma province (1968 paper) suggests values of $p_m = 0.821$ and $p_f = 0.727$, whose product is 0.597, if one assumes a very narrow marriage range, of radius 1.6 km. This includes only a small village, of birth or residence. With a 4.1-km radius for the marriage range, which often includes the nearest village, the values are $p_m = 0.847$, $p_f = 0.771$ (product 0.653). These data suggest $P = p_m \times p_f$ values between 0.6 and 0.7. But there is the possibility that other types of migration, such as change of social or economic status, could add their effect, as well as breaks of friendly relations, which are not rare among relatives, but for which it may be difficult to find measurements. These would further decrease the value of $P$. We have detected in section 4.3 a complication that goes counter to the progressive loss of contact between relatives with increasing remoteness of the relationship: the help given by women to maintaining family ties. But we believe that this factor has a lesser effect than the factors we are investigating at present.

TABLE 4.7. Comparison of observed and expected frequencies of nine consanguinity degrees for different values of $P = p_m \times p_f$

| | Observed | | Expected % | | | | |
|---|---|---|---|---|---|---|---|
| Degree | No. | % | $P = 0.7$ | $P = 0.6$ | $P = 0.5$ | $P = 0.4$ | Considering avoidance |
| 1. Uncle–niece | 28 | 0.91 | 0.77 | 1.3 | 2.2 | 3.9 | 0.63 |
| 2. Aunt–nephew | 3 | 0.097 | 0.096 | 0.16 | 0.28 | 0.49 | 0.081 |
| 3. First cousins | 740 | 24.0 | 14.9 | 21.5 | 31.0 | 43.4 | 24.7 |
| 4. 1½ HSB | 198 | 6.4 | 4.0 | 5.0 | 5.9 | 6.7 | 6.8 |
| 5. 1½ WSB | 49 | 1.6 | 1.3 | 1.6 | 1.9 | 2.1 | 2.1 |
| 6. Second cousins | 1,072 | 34.8 | 23.6 | 25.2 | 25.2 | 22.6 | 31.2 |
| 7. 2½ HSB | 314 | 10.2 | 10.6 | 9.7 | 8.1 | 5.8 | 9.2 |
| 8. 2½ WSB | 106 | 3.4 | 4.8 | 4.4 | 3.6 | 2.6 | 4.1 |
| 9. Third cousins | 569 | 18.5 | 40.0 | 31.3 | 21.7 | 12.5 | 21.0 |
| Total | 3,079 | 99.9 | | | | | |

*Note.* HSB, husband in the shorter branch; WSB, wife in the shorter branch.

For these reasons we have not experimented widely on changing the demographic quantities, which seem better ascertained and more stable. Their change might bring about only a small advantage in fitting. We have tried instead a relatively large range of $P$ values, from $P = 0.7$ to $P = 0.4$, shown in table 4.7. Our observations on geographic migration would point only to the upper part of the range, near 0.7, as being realistic. But we know that other possible causes can simulate geographic migration and add to the $P$ value. The range of explanations covered by the four values of $P$ in the table helps us by showing why a theory based exclusively on age and migration cannot explain these data.

The observations show a highly significant peak of relative frequency of second cousins, and in table 4.7 one can see that even a fairly wide variation of $P$ does not allow a relative peak of second cousins in theoretical expectations as high as the observed one. With little migration ($P = .7$) the frequency of even cousins increases geometrically with degree. With more migration and therefore smaller $P$ the frequency of even cousins eventually inverts its trend and decreases geometrically as the degree of the cousins increases.

There are three possible explanations of this situation and all involve an external factor and could generate the observed anomaly we are trying to explain, the relative peak of second cousins:

1. There is in this population a trend to avoid close consanguineous marriage, probably for fear of genetic damage or of effectiveness of prohibition.
2. Third cousins, or, more generally, remote consanguinities, are easily undetected.
3. As we show in chapter 6, it is likely that most multiple consanguinities are not reported as such, but that only the closest consanguineous ties are reported, which may distort the data.

In generating a function to evaluate avoidance, the function should clearly decrease from 100% to nearly 0% as the consanguinity degree increases. Taking $b$ as degree, and putting $z = b^2/C$, one such function is

$$A = e^{-z} \tag{4.11}$$

TABLE 4.8. Observed and expected frequency of consanguineous marriages in the Parma diocese

| Degree | b | $n_c$ | $m_i$ | $m_j$ | $f_i$ | $f_j$ | Observed | $f_c$ | $f_{age}$ | $f_{mig}$ | $1 - avd$ | $f_c \times K$ | Expected | $\chi^2$ | Observed/expected |
|--------|-----|------|------|------|------|------|----------|---------|---------|---------|-------|---------|--------|--------|------|
| 1 | 1.0 | 2.0 | 0.0 | 0.5 | 0.0 | 0.5 | 28 | 0.00016 | 0.00207 | 0.40000 | 0.095 | 0.00646 | 19.90 | 3.30 | 1.41 |
| 2 | 1.0 | 2.0 | 0.5 | 0.0 | 0.5 | 0.0 | 3 | 0.00002 | 0.00026 | 0.40000 | 0.095 | 0.00081 | 2.49 | 0.10 | 1.21 |
| 3 | 2.0 | 4.0 | 0.5 | 0.5 | 0.5 | 0.5 | 740 | 0.00631 | 0.02861 | 0.16000 | 0.330 | 0.24790 | 763.28 | 0.71 | 0.97 |
| 4 | 3.0 | 8.0 | 0.5 | 1.0 | 0.5 | 1.0 | 198 | 0.00167 | 0.00549 | 0.06400 | 0.593 | 0.06850 | 210.92 | 0.79 | 0.94 |
| 5 | 3.0 | 8.0 | 1.0 | 0.5 | 1.0 | 0.5 | 49 | 0.00053 | 0.00174 | 0.06400 | 0.593 | 0.02170 | 66.82 | 4.75 | 0.73 |
| 6 | 4.0 | 16.0 | 1.0 | 1.0 | 1.0 | 1.0 | 1,072 | 0.00759 | 0.02321 | 0.02560 | 0.798 | 0.31166 | 959.61 | 13.16 | 1.12 |
| 7 | 5.0 | 32.0 | 1.0 | 1.5 | 1.0 | 1.5 | 314 | 0.00223 | 0.00742 | 0.01024 | 0.918 | 0.09170 | 282.33 | 3.55 | 1.11 |
| 8 | 5.0 | 32.0 | 1.5 | 1.0 | 1.5 | 1.0 | 106 | 0.00101 | 0.00335 | 0.01024 | 0.918 | 0.04143 | 127.57 | 3.65 | 0.83 |
| 9 | 6.0 | 64.0 | 1.5 | 1.5 | 1.5 | 1.5 | 569 | 0.00511 | 0.02004 | 0.00409 | 0.973 | 0.20983 | 646.07 | 9.19 | 0.88 |
| Total | | | | | | | 3,079 | | | | | | | 39.21 | |

Note. The expected frequency is subdivided into three factors, measuring respectively the probability of each degree due to age (see formula 4.9 and the demographic quantities of table 4.1), the loss due to migration (see formula 4.10 with $p_m \times p_f = 0.40$), and that due to avoidance (avd) of close consanguineous marriages (see formula 4.12 with $C = 10.0$). $K = 1/\text{sum of expectations for all degrees}$.

where $C$ is a constant to be fitted numerically. The acceptance function is $1 - A$, or

$$f(b) = 1 - e^{-z} \qquad (4.12)$$

by which $f_c(b)$ (see equation 4.8) must be multiplied. Fitting to the observed values of table 4.7 the function that is the product of (4.8) and (4.12), the approximate $P$ and $C$ values were obtained, minimizing chi square

$$C = 10; \quad P = .40$$

Table 4.8 shows the observed frequencies of the nine degrees in Parma, their corresponding expected frequencies, obtained by fitting the loss of consanguineous marriages due to avoidance for each degree, and the total avoidance frequency. Chi square is large compared with standard significance levels ($\chi^2 = 39.20$; 8 df, $P < 0.001$), but one cannot expect a perfect fit because of the heterogeneity of the data in space and time (section 4.5). In fact, considering that the majority of the parameters employed for the demographic part are not the result of fitting but are observed independently, the fit looks surprisingly good.

CHAPTER 5

# Consanguinity, Inbreeding, and Observed Genetic Drift in the Parma Valley

## 5.1 THE PARMA VALLEY AND THE ORIGIN OF THIS INVESTIGATION

This research originated in the early 1950s and, although some parts of it were published (Cavalli-Sforza 1969, Cavalli-Sforza and Bodmer 1971, 1999, Cavalli-Sforza and Feldman 1990), a complete report never appeared. In fact, it is only very recently that a final evaluation of the results has been made. The project began in 1954, with the idea of collecting data on the genetic diversity of the population inhabiting the villages of a very diverse area, that of the valley of Parma, in which one could distinguish essentially three types of habitat: the towns of the plains and the villages of the hills and the mountains. There is no clear-cut genetic or cultural discontinuity among the three environments, but there is a continuous change in economic conditions from low to high altitude. Towns in the plains and hills are blessed with a rich agriculture and florid economy, while in the mountains living conditions are marginal. Only recently has there been some economic support from tourism. While the plains and to some extent the lower hills are among the richest lands in northern Italy, in the mountains many people were still accustomed, at the time this inquiry was begun, to collecting chestnuts in the woods to secure an acceptable calorie input in the worst months of the year. Not surprisingly, there is a ten-fold difference in population density among sections of the area.

The gathering of the material was from the beginning divided in two parts: the collection of demographic data from parish books and consanguinity dispensations and the collection of genetic data from a population sample. For the latter purpose, blood samples were collected from a total of 2,815 individuals in 74 villages. At that time only three blood group systems (ABO, MN, RH) were really highly informative. As described in section 1.7, the aim was to use the demographic data to make an estimate of the genetic variation to be expected among villages in the various sections of the valley on the basis of random genetic drift; to use the blood group data to estimate the observed genetic variation among villages; and to compare the former with the latter. At the time, the importance and the very existence of random genetic drift was considered questionable, or at least trivial compared with that of natural selection.

Had we been able to conserve the samples for later use, we could have examined other genetic systems. This was not the case, because of a failure of the refrigerator in which the blood samples were stored. Much later it became clear to us that surnames could supply an excellent alternative to standard genetic data. We studied them in other parts of Italy, and quite recently we returned to the Parma valley for a supplementary investigation.

We examined and compared the demographic and genetic variation of the valley in different times and with a variety of methods, and results are given in this chapter. Until recently we were not entirely satisfied by the results of the comparison with the existing methods for predicting variation due to drift, as shown by the analysis made in 1971 in the book by Cavalli-Sforza and Bodmer. Later study showed that a method we had started in 1963 for the analysis of drift in the most significant part of the valley, using a computer simulation (Cavalli-Sforza and Zei 1967), gave, when properly interpreted, the most satisfactory answer. Those results were not published before except in small part and are given in chapter 6.

## 5.2 GEOGRAPHY OF THE PARMA VALLEY

Parma is a torrent originating in the Apennine and descending in a northeastern direction until it flows into the Po, the longest Italian river, the basin of which forms the greatest part of northern Italy. Not far from its end, the Parma River goes through the city by the same name, the capital of a duchy until 1860, with about 200,000 inhabitants in 1970. The valley is defined by the river, and is 70 km long as the crow flies. The greatest width is 20 km and the shape is approximately rectangular. The city of Parma is on the river, about 20 km south of its confluence into the Po River. The altitude grows regularly toward the south up to a maximum of 1861 meters on the crest of the Apennines; south of it and on the opposite part of the mountain chain is a valley leading to Pontremoli, a town of Tuscany. There are roads across the Apennines only east and west of the Parma valley, which is therefore fairly isolated from Tuscany, but there are reasonable connections between the Parma valley and the two parallel, adjacent valleys, that of the river Baganza to the west and that of the river Enza to the east. The Baganza valley belongs to the Diocese (the religious administrative unit) and the Province (Italian government administrative unit) of Parma, while the Enza River is the boundary with the Diocese and Province of Reggio Emilia, east of Parma. Figure 5.1 shows the three major rivers, the roads, the names of the *comuni* (communes) (the smallest administrative units), and those of the parishes. Table 5.1 shows the number of inhabitants of the *comuni* and its variation in the interval 1861–1961. One observes that the total number of inhabitants increased but only in the city or nearby; in the four mountain communes (the last four of table 5.1), and to some extent in the hills, the population has been decreasing since the 1920s.

Our investigation has been limited to the parishes of the Parma valley south of the city. There are 12 *comuni* in the valley, in which we sampled at least one parish, and there are, or were at the beginning of our investigation, 74 parishes (excluding those of the city of Parma, or further north). Parishes correspond to inhabited villages. Almost all parishes of the Parma valley already existed in a survey of the eleventh century. Parish books were established in most parishes after the Council of Trento (1545–1562) with the obligation to register births (i.e., baptisms), deaths, and marriages. Marriages are, as a rule, registered in the parish of the bride. Some parishes began later, so their books begin later. Registrations followed a routine format that very early began to be standard, but surnames are rare and relatively unstable in the very first years after the Trento Council. The

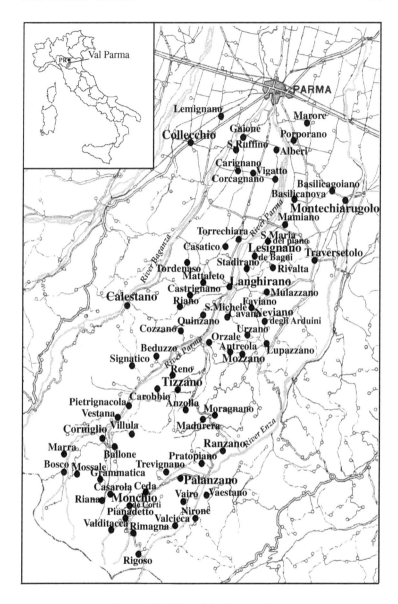

FIGURE 5.1. Map of the Parma valley.

TABLE 5.1. Number of inhabitants in the communes of the Parma valley from official censuses of the Italian government

| Commune | 1861 | 1871 | 1881 | 1901 | 1911 | 1921 | 1931 | 1936 | 1951 | 1961 |
|---|---|---|---|---|---|---|---|---|---|---|
| Collecchio | 2,693 | 4,158 | 4,553 | 5,782 | 6,803 | 8,066 | 8,149 | 8,400 | 9,035 | 8,851 |
| Parma[a] | 44,126 | 45,511 | 43,553 | 48,523 | 50,725 | 56,685 | 68,713 | 71,858 | 122,978 | 141,203 |
| Montechiarugolo | 4,055 | 4,244 | 4,485 | 4,699 | 5,360 | 6,253 | 6,527 | 6,485 | 6,802 | 6,082 |
| Traversetolo | 3,634 | 3,842 | 4,257 | 4,476 | 5,100 | 6,075 | 6,279 | 6,147 | 5,975 | 5,750 |
| Langhirano | 5,502 | 5,731 | 6,298 | 7,155 | 8,043 | 8,410 | 8,045 | 7,752 | 7,209 | 6,554 |
| Lesignano de' Bagni | 2,931 | 3,006 | 3,452 | 3,879 | 4,481 | 4,449 | 4,462 | 4,176 | 3,687 | 2,990 |
| Calestano | 3,051 | 2,961 | 3,603 | 3,331 | 3,676 | 4,011 | 3,691 | 3,468 | 3,244 | 2,585 |
| Neviano Arduini | 5,580 | 5,647 | 6,373 | 7,429 | 8,258 | 8,806 | 9,425 | 8,873 | 8,164 | 5,995 |
| Tizzano | 3,489 | 3,308 | 4,166 | 4,999 | 5,252 | 5,665 | 5,096 | 4,730 | 4,488 | 3,732 |
| Corniglio | 5,533 | 5,080 | 6,204 | 7,157 | 7,401 | 7,659 | 7,174 | 6,608 | 6,112 | 5,139 |
| Palanzano | 2,271 | 2,231 | 2,735 | 3,102 | 3,495 | 4,038 | 3,943 | 3,586 | 3,463 | 2,710 |
| Monchio | 2,327 | 2,110 | 2,906 | 3,480 | 3,659 | 3,855 | 3,934 | 3,333 | 3,306 | 2,785 |
| Totals | 85,192 | 87,829 | 92,585 | 104,012 | 112,253 | 123,972 | 135,438 | 135,416 | 184,463 | 194,376 |

[a]City parishes were excluded from count; inhabitants were included.

majority of surnames in the upper part of the valley began at the end of the sixteenth century. Some city parishes have maintained books of baptisms since the fifteenth century, but these were not examined by us.

Only one bridge across the Parma River south of the city of Parma was made by Romans, but there were most probably numerous places of crossing, as shown in an 1820 map by mulepaths that led down to the river, and continued at exactly the same place on the opposite side. That the river was no major impediment to local communication is shown by an analysis of the frequency of marriages across and on the same side of the river (Crumpacker et al. 1976): the analysis of distances by air and on shortest mulepaths showed a very high correlation, indicating reasonably good communication throughout most of the valley. Communication with the Tuscan region located south of Parma, across the Apennines, was not easy from the Parma valley, but was easier in parallel valleys, east and west.

## 5.3 CONSANGUINITY AND INBREEDING IN THE PARMA VALLEY

The 74 parishes of the valley are listed in table 5.2, together with the number of inhabitants (from the 1951 census), their altitude, the population density of the commune to which they belong, the number of marriages in the period 1851–1950, taken from parish books, and the numbers of consanguineous marriages listed in the parish books. Relative frequencies of consanguineous marriages and the usual average inbreeding were calculated in two ways: using all available marriages and excluding second cousins once removed, third cousins, and multiple consanguinity, which are missing in later years.

The *comune* of Collecchio has only one parish (Lemignano) in the Parma valley, which has only one consanguineous marriage. In the *comune* of Parma, five parishes (out of eight) have very low numbers of consanguineous marriages. Because there was a chance of some inaccuracy of these records, we left out in the correlation analysis reported later all parishes with less than eight consanguineous marriages, and, therefore, the *comune* of Collecchio.

Consanguinity increases clearly with altitude, as shown by a sim-

TABLE 5.2. Demographic data and consanguinity data of 74 parishes of the Parma valley, 1851–1950

| Commune | Density (inhabitants/ km²) | Parish | Altitude | Inhabitants (1951) | Total marriages (1851– 1950) |
|---|---|---|---|---|---|
| Parma | 472 | Alberi | 89 | 638 | 315 |
| | | Carignano | 130 | 1,612 | 792 |
| | | Corcagnano | 121 | 693 | 436 |
| | | Gaione | 98 | 667 | 389 |
| | | Marore | 61 | 484 | 338 |
| | | Porporano | 86 | 812 | 496 |
| | | San Ruffino | 112 | 417 | 272 |
| | | Vigatto | 113 | 1,182 | 934 |
| | | *Total* | | *6,505* | *3,972* |
| Montechiarugolo | 142 | Bas. Goiano | 119 | 1,733 | 1,092 |
| | | Bas. Nova | 131 | 2,466 | 1,627 |
| | | *Total* | | *4,199* | *2,719* |
| Traversetolo | 109 | Bannone | 160 | 612 | 478 |
| | | Mamiano | 158 | 732 | 606 |
| | | *Total* | | *1,344* | *1,084* |
| Langhirano | 102 | Arola | 199 | 666 | 1,473 |
| | | Casatico | 384 | 340 | 305 |
| | | Castrignano | 555 | 692 | 803 |
| | | Cozzano | 704 | 592 | 426 |
| | | Mattaleto | 337 | 332 | 177 |
| | | Quinzano | 527 | 164 | 126 |
| | | Riano | 608 | 255 | 204 |
| | | Strognano | 456 | 251 | 159 |
| | | Tordenaso | 450 | 319 | 281 |
| | | Torrechiara | 221 | 506 | 321 |
| | | *Total* | | *4,117* | *4,275* |
| Lesignano de' Bagni | 78 | Cavana | 465 | 560 | 501 |
| | | Faviano | 392 | 344 | 311 |
| | | Lesignano | 252 | 602 | 420 |
| | | Mulazzano | 440 | 868 | 551 |
| | | Rivalta | 357 | 267 | 202 |
| | | S.M. Del Piano | 204 | 723 | 557 |
| | | Stadirano | 338 | 323 | 239 |
| | | *Total* | | *3,687* | *2,781* |
| Calestano | 57 | Calestano | 417 | 1,045 | 915 |
| Neviano Arduini | 77 | Antreola | 700 | 265 | 135 |
| | | Lupazzano | 532 | 343 | 203 |

| | | Type of cousins | | | | | | | % consanguineous/ | α × 1,000 | |
|---|---|---|---|---|---|---|---|---|---|---|---|
| 12 | 22 | 23 | 33 | 34 | 44 | Half | Multiple | Total | total | Total | 12–33 |
| 0 | 1 | 1 | 1 | 0 | 0 | 0 | 0 | 3 | 0.95 | 0.347 | 0.347 |
| 0 | 3 | 1 | 4 | 2 | 0 | 0 | 2 | 12 | 1.52 | 0.454 | 0.355 |
| 0 | 0 | 1 | 0 | 0 | 0 | 0 | 0 | 1 | 0.23 | 0.072 | 0.072 |
| 0 | 1 | 0 | 1 | 0 | 1 | 0 | 0 | 3 | 0.77 | 0.211 | 0.201 |
| 0 | 1 | 0 | 1 | 0 | 0 | 0 | 0 | 2 | 0.59 | 0.231 | 0.231 |
| 1 | 4 | 1 | 5 | 0 | 0 | 0 | 0 | 11 | 2.22 | 0.977 | 0.977 |
| 0 | 1 | 0 | 0 | 0 | 0 | 0 | 0 | 1 | 0.37 | 0.230 | 0.230 |
| 0 | 2 | 1 | 2 | 1 | 1 | 1 | 0 | 8 | 0.86 | 0.230 | 0.201 |
| *1* | *13* | *5* | *14* | *3* | *2* | *1* | *2* | *41* | *1.03* | *0.358* | *0.330* |
| 0 | 6 | 1 | 4 | 0 | 0 | 0 | 0 | 11 | 1.01 | 0.429 | 0.429 |
| 1 | 12 | 1 | 8 | 2 | 0 | 0 | 0 | 24 | 1.48 | 0.643 | 0.643 |
| *1* | *18* | *2* | *12* | *2* | *0* | *0* | *0* | *35* | *1.29* | *0.557* | *0.552* |
| 1 | 3 | 1 | 4 | 3 | 1 | 0 | 0 | 13 | 2.72 | 0.907 | 0.850 |
| 0 | 3 | 4 | 2 | 0 | 0 | 0 | 0 | 9 | 1.49 | 0.567 | 0.567 |
| *1* | *6* | *5* | *6* | *3* | *1* | *0* | *0* | *22* | *2.03* | *0.717* | *0.692* |
| 0 | 11 | 1 | 6 | 2 | 0 | 0 | 0 | 20 | 1.36 | 0.562 | 0.552 |
| 0 | 5 | 0 | 3 | 1 | 1 | 0 | 0 | 10 | 3.28 | 1.217 | 1.178 |
| 0 | 4 | 2 | 22 | 0 | 2 | 0 | 1 | 31 | 3.86 | 0.866 | 0.817 |
| 0 | 6 | 5 | 21 | 13 | 7 | 0 | 5 | 57 | 13.38 | 2.760 | 2.017 |
| 1 | 9 | 2 | 7 | 0 | 2 | 0 | 0 | 21 | 11.86 | 4.899 | 4.855 |
| 0 | 2 | 2 | 3 | 0 | 0 | 0 | 0 | 7 | 5.56 | 1.860 | 1.860 |
| 0 | 5 | 2 | 11 | 3 | 5 | 0 | 0 | 29 | 14.22 | 3.657 | 2.681 |
| 0 | 1 | 0 | 0 | 0 | 3 | 0 | 0 | 4 | 2.52 | 0.467 | 0.393 |
| 0 | 3 | 1 | 3 | 0 | 3 | 0 | 0 | 10 | 3.56 | 0.987 | 0.945 |
| 0 | 1 | 0 | 3 | 0 | 1 | 0 | 0 | 5 | 1.56 | 0.353 | 0.341 |
| *1* | *47* | *15* | *79* | *19* | *24* | *0* | *9* | *194* | *4.54* | *1.259* | *1.115* |
| 0 | 5 | 2 | 12 | 1 | 0 | 0 | 2 | 22 | 4.39 | 1.325 | 1.123 |
| 0 | 5 | 0 | 2 | 0 | 1 | 0 | 0 | 8 | 2.57 | 1.118 | 1.105 |
| 0 | 3 | 0 | 8 | 1 | 5 | 0 | 0 | 17 | 4.05 | 0.809 | 0.744 |
| 0 | 7 | 2 | 20 | 3 | 6 | 0 | 2 | 40 | 7.26 | 1.701 | 1.475 |
| 0 | 3 | 0 | 5 | 0 | 2 | 0 | 0 | 10 | 4.95 | 1.354 | 1.315 |
| 0 | 4 | 3 | 3 | 3 | 1 | 0 | 1 | 15 | 2.69 | 0.863 | 0.701 |
| 0 | 2 | 1 | 6 | 5 | 4 | 0 | 0 | 18 | 7.53 | 1.275 | 1.046 |
| *0* | *29* | *8* | *56* | *13* | *19* | *0* | *5* | *130* | *4.67* | *1.204* | *1.056* |
| 0 | 14 | 8 | 22 | 8 | 11 | 0 | 4 | 67 | 7.32 | 1.780 | 1.605 |
| 0 | 3 | 4 | 10 | 0 | 0 | 0 | 0 | 17 | 12.59 | 3.472 | 3.472 |
| 0 | 2 | 2 | 4 | 3 | 0 | 0 | 0 | 11 | 5.42 | 1.347 | 1.232 |

TABLE 5.2. *Continued*

| Commune | Density (inhabitants/ km²) | Parish | Altitude | Inhabitants (1951) | Total marriages (1851– 1950) |
|---|---|---|---|---|---|
| | | Mozzano | 564 | 330 | 243 |
| | | Neviano Ard. | 514 | 1,187 | 684 |
| | | Orzale | 421 | 242 | 213 |
| | | Urzano | 460 | 517 | 460 |
| | | *Total* | | *2,884* | *1,938* |
| Tizzano | 57 | Anzolla | 713 | 194 | 199 |
| | | Carrobbio | 691 | 290 | 208 |
| | | Cereto | 590 | 289 | 316 |
| | | Madurera | 774 | 271 | 240 |
| | | Moragnano | 772 | 411 | 261 |
| | | Musiara Inf. | 889 | 185 | 308 |
| | | Reno | 522 | 523 | 364 |
| | | Tizzano | 814 | 800 | 564 |
| | | *Total* | | *2,963* | *2,460* |
| Corniglio | 37 | Ballone | 823 | 426 | 381 |
| | | Beduzzo | 584 | 1,084 | 679 |
| | | Bosco | 841 | 639 | 456 |
| | | Corniglio | 690 | 887 | 600 |
| | | Grammatica | 1026 | 151 | 115 |
| | | Marra | 852 | 208 | 270 |
| | | Mossale | 909 | 171 | 176 |
| | | Petrignacola | 634 | 437 | 349 |
| | | Signatico | 816 | 220 | 182 |
| | | Vestana | 801 | 261 | 484 |
| | | Villula | 678 | 164 | 123 |
| | | *Total* | | *4,648* | *3,815* |
| Palanzano | 49 | Nirone | 720 | 200 | 127 |
| | | Palanzano | 691 | 660 | 593 |
| | | Pratopiano | 713 | 293 | 174 |
| | | Ranzano | 609 | 587 | 375 |
| | | Trevignano | 680 | 131 | 136 |
| | | Vaestano | 642 | 209 | 221 |
| | | Vairo | 840 | 221 | 165 |
| | | Valcieca | 893 | 171 | 126 |
| | | *Total* | | *2,472* | *1,917* |
| Monchio | 48 | Casarola | 996 | 193 | 184 |
| | | Ceda | 806 | 251 | 159 |
| | | Lugagnano | 742 | 473 | 384 |

| | | | Type of cousins | | | | | | % consanguineous/ | α × 1,000 | |
|---|---|---|---|---|---|---|---|---|---|---|---|
| 12 | 22 | 23 | 33 | 34 | 44 | Half | Mul-tiple | Total | total | Total | 12–33 |
| 0 | 1 | 3 | 4 | 1 | 0 | 0 | 2 | 11 | 4.53 | 1.961 | 0.900 |
| 0 | 10 | 1 | 12 | 1 | 4 | 0 | 5 | 33 | 4.82 | 1.656 | 1.234 |
| 0 | 2 | 0 | 5 | 0 | 0 | 0 | 1 | 8 | 3.76 | 1.100 | 0.954 |
| 0 | 6 | 4 | 7 | 4 | 3 | 0 | 2 | 26 | 5.65 | 1.486 | 1.325 |
| *0* | *24* | *14* | *42* | *9* | *7* | *0* | *10* | *106* | *5.47* | *1.687* | *1.338* |
| 0 | 3 | 1 | 5 | 0 | 2 | 0 | 0 | 11 | 5.53 | 1.531 | 1.492 |
| 0 | 7 | 2 | 6 | 6 | 8 | 0 | 1 | 30 | 14.42 | 3.380 | 2.855 |
| 0 | 2 | 0 | 4 | 1 | 2 | 0 | 0 | 9 | 2.85 | 0.643 | 0.593 |
| 0 | 4 | 1 | 6 | 0 | 0 | 0 | 0 | 11 | 4.58 | 1.563 | 1.563 |
| 0 | 3 | 4 | 6 | 0 | 0 | 1 | 0 | 14 | 5.36 | 1.616 | 1.557 |
| 0 | 2 | 5 | 11 | 2 | 2 | 0 | 0 | 22 | 7.14 | 1.547 | 1.471 |
| 0 | 4 | 3 | 10 | 0 | 0 | 0 | 1 | 18 | 4.95 | 1.545 | 1.374 |
| 0 | 11 | 2 | 14 | 3 | 6 | 0 | 2 | 37 | 6.56 | 1.828 | 1.607 |
| *0* | *35* | *18* | *62* | *12* | *20* | *1* | *4* | *152* | *6.18* | *1.658* | *1.512* |
| 0 | 9 | 2 | 26 | 14 | 15 | 0 | 2 | 68 | 17.85 | 3.209 | 2.707 |
| 0 | 12 | 5 | 31 | 6 | 12 | 2 | 1 | 69 | 10.16 | 2.347 | 2.048 |
| 2 | 3 | 4 | 23 | 8 | 6 | 0 | 3 | 49 | 10.75 | 2.450 | 2.022 |
| 0 | 22 | 8 | 33 | 1 | 10 | 0 | 1 | 75 | 12.50 | 3.750 | 3.568 |
| 0 | 11 | 4 | 19 | 5 | 3 | 0 | 6 | 48 | 41.74 | 11.447 | 9.647 |
| 0 | 11 | 5 | 14 | 5 | 9 | 0 | 0 | 44 | 16.30 | 4.210 | 3.935 |
| 0 | 0 | 6 | 19 | 0 | 3 | 0 | 0 | 28 | 15.91 | 2.819 | 2.752 |
| 0 | 6 | 2 | 19 | 1 | 3 | 0 | 2 | 33 | 9.46 | 2.608 | 2.104 |
| 0 | 2 | 1 | 12 | 0 | 6 | 0 | 1 | 22 | 12.09 | 2.103 | 1.889 |
| 0 | 8 | 2 | 12 | 11 | 7 | 0 | 1 | 40 | 8.26 | 1.784 | 1.550 |
| 1 | 8 | 0 | 1 | 0 | 1 | 0 | 0 | 11 | 8.94 | 5.240 | 5.208 |
| *3* | *92* | *39* | *209* | *51* | *75* | *2* | *16* | *487* | *12.77* | *3.128* | *2.781* |
| 0 | 6 | 2 | 9 | 0 | 5 | 0 | 1 | 23 | 18.11 | 5.198 | 4.552 |
| 0 | 24 | 9 | 38 | 6 | 6 | 0 | 11 | 94 | 15.85 | 5.099 | 4.005 |
| 0 | 3 | 0 | 5 | 6 | 2 | 0 | 1 | 17 | 9.77 | 1.886 | 1.527 |
| 0 | 11 | 2 | 17 | 2 | 7 | 0 | 4 | 43 | 11.47 | 2.990 | 2.708 |
| 0 | 4 | 2 | 9 | 3 | 6 | 0 | 0 | 24 | 17.65 | 3.676 | 3.332 |
| 0 | 5 | 1 | 7 | 1 | 3 | 0 | 1 | 18 | 8.14 | 2.422 | 2.050 |
| 0 | 4 | 0 | 4 | 1 | 0 | 0 | 0 | 9 | 5.45 | 1.941 | 1.894 |
| 1 | 4 | 3 | 7 | 0 | 0 | 0 | 2 | 16 | 12.70 | 4.836 | 3.596 |
| *0* | *61* | *19* | *96* | *19* | *29* | *0* | *20* | *244* | *12.73* | *3.702* | *3.081* |
| 0 | 6 | 7 | 18 | 1 | 0 | 0 | 5 | 37 | 20.11 | 6.072 | 4.755 |
| 0 | 3 | 2 | 8 | 3 | 5 | 0 | 0 | 21 | 13.21 | 2.629 | 2.358 |
| 1 | 16 | 5 | 38 | 12 | 18 | 0 | 6 | 96 | 25.00 | 5.900 | 4.883 |

TABLE 5.2. *Continued*

| Commune | Density (inhabitants/ km²) | Parish | Altitude | Inhabitants (1951) | Total marriages (1851– 1950) |
|---------|---------------------------|--------|----------|--------------------|-----------------------------|
|         |                           | Monchio | 820 | 600 | 444 |
|         |                           | Pianadetto | 975 | 274 | 235 |
|         |                           | Riana | 1027 | 143 | 131 |
|         |                           | Rigoso | 1131 | 267 | 169 |
|         |                           | Rimagna | 963 | 275 | 185 |
|         |                           | Trefiumi | 938 | 345 | 233 |
|         |                           | Valditacca | 1010 | 346 | 263 |
|         |                           | *Total* |  | *3,167* | *2,387* |
|         |                           | **General Total** |  | **37,031** | **28,263** |

ple inspection of table 5.2. The total percentage of consanguineous marriages increases from less than 1% in the plains, at about 100 meters above sea level, to a maximum of 42% in the parishes at the greatest altitude (around 1000 meters). Correlations with altitude, the number of inhabitants per parish, and population density are shown in table 5.3 for four quantities defining different aspects of consanguinity calculated from the data in table 5.2. Three of them are percentages of different types of consanguinity, and the fourth is the average inbreeding coefficient. Altitude seems to be the best predictor of consanguinity and the number of inhabitants per parish the worst; population density is intermediate as judged from the correlation values, all of which, however, are highly significant. Of the four consanguinity/inbreeding parameters, correlations with first cousins are a little inferior to the others. The difference is not large, and in the case of the number of inhabitants it is less sharp. The correlations between these four quantities are shown in table 5.4.

Given the importance of altitude we have grouped the data in two classes, plains + hills (P + H) and mountain (M), and show in table 5.5 the averages and standard errors of four demographic variables and of the four consanguinity quantities. The differences are all highly significant. In the mountains the percentage of consanguinity

| | | Type of cousins | | | | | | | % consanguineous/ | α × 1,000 | |
| 12 | 22 | 23 | 33 | 34 | 44 | Half | Mul-tiple | Total | total | Total | 12–33 |
|---|---|---|---|---|---|---|---|---|---|---|---|
| 0 | 17 | 4 | 39 | 7 | 16 | 0 | 3 | 86 | 19.37 | 4.381 | 4.047 |
| 0 | 11 | 1 | 36 | 3 | 8 | 0 | 1 | 60 | 25.53 | 5.718 | 5.452 |
| 0 | 8 | 1 | 11 | 2 | 1 | 0 | 1 | 24 | 18.32 | 5.755 | 5.367 |
| 0 | 5 | 5 | 19 | 3 | 5 | 0 | 0 | 37 | 21.89 | 4.785 | 4.530 |
| 0 | 13 | 7 | 34 | 3 | 2 | 0 | 2 | 61 | 32.97 | 9.122 | 8.446 |
| 0 | 13 | 5 | 41 | 6 | 8 | 0 | 3 | 76 | 32.62 | 7.645 | 6.907 |
| 1 | 8 | 4 | 28 | 6 | 6 | 0 | 5 | 58 | 22.05 | 5.822 | 4.515 |
| *2* | *100* | *41* | *272* | *46* | *69* | *0* | *26* | *556* | *23.29* | *5.719* | *5.040* |
| *9* | *439* | *174* | *870* | *185* | *257* | *4* | *96* | *2,034* | *7.20* | *1.914* | *1.684* |

is on average 3 times higher than that in the plains + hills, while the inbreeding coefficient is 2.7 times higher.

In the next sections we compare these quantities with those obtained for genetic drift in the same set of data.

## 5.4 BLOOD GROUPS AND GENETIC DRIFT

Blood samples were obtained from all parishes listed in table 5.2, which, with very few exceptions, have a one-to-one correspondence with the villages present in the area. Numbers of individuals in the sample from each village varied from 16 to 101, but the great majority were between 25 and 45. Table 5.6 shows the phenotype frequencies in all 74 parishes of the three blood group systems investigated, which were the genes most easily testable at the time the analysis was done (1954–1958). Gene frequencies calculated by maximum likelihood are given in table 5.7.

The great majority of individuals sampled were less related than parent–child or sib–sib, but there were some exceptions. This generates a redundancy of genetic information, which may affect the variance of the estimate of gene frequencies. There is no bias in the estimate of the gene frequency but the error variance is underesti-

TABLE 5.3. Correlation of consanguinity data with demography of the
Parma valley

|  | Altitude | Number of inhabitants | Density (inhabitants/km$^2$) |
|---|---|---|---|
| Consanguineous/total marriages | 0.846 (<0.001) | −0.558 (<0.001) | −0.722 (<0.001) |
| First cousins/total marriages | 0.725 (<0.001) | −0.525 (<0.001) | −0.660 (<0.001) |
| Second cousins/total marriages | 0.854 (<0.001) | −0.514 (<0.001) | −0.686 (<0.001) |
| Average inbreeding coefficient α | 0.834 (<0.001) | −0.556 (<0.001) | −0.687 (<0.001) |

*Note.* P values in parentheses.

mated for dominant pairs of alleles. This problem can be circum-
vented by a method devised and tested by simulation (Guglielmino-
Matessi and Zei 1979), if one calculates a *coefficient of redundancy*,
which indicates how to reduce the number of individuals to correct
for redundancy. The coefficient is calculated from the number of
first-degree relatives in the sample by subtracting from the total number of chromosomes tested (twice the number of individuals), one if a
parent and child are included in the sample, one and a half for one
parent and two children, one and three-quarters for one parent and
three children, and using slightly more complicated formula for sibs
or other more complicated situations, as indicated in the original pa-

TABLE 5.4. Parma valley: correlations between consanguinity data in 65
parishes with more than 7 consanguineous marriages

|  | First cousins/ total marriages | Second cousins/ total marriages | Average inbreeding coefficient α |
|---|---|---|---|
| % consanguineous marriages/total marriages | 0.824 | 0.953 | 0.950 |
| % first cousins/total marriages |  | 0.708 | 0.928 |
| % second cousins/total marriages |  |  | 0.880 |

TABLE 5.5. Parma valley: average values of ecological/demographic and consanguinity/inbreeding values in plains + hills and mountains

| Type of data | Plains + Hills | Mountains | Total | P + H vs. M t test |
|---|---|---|---|---|
| Ecological/demo-graphic | | | | |
| Altitude (meters above sea level) | 323 ± 31.8 (61 − 704) | 800 ± 23.8 (522 − 1131) | 561 ± 34.2 (61 − 1131) | |
| Population density ($N$/km$^2$) | 176 ± 26.19 (57 − 472) | 47 ± 1.2 (37 − 57) | 111 ± 15.0 (37 − 472) | 4.93*** |
| No. of inhabitants/village | 657 ± 77.9 (164 − 2,466) | 358 ± 37.2 (131 − 1,084) | 508 ± 46.3 (131 − 2,466) | 3.46*** |
| No. of parishes | 37 | 37 | 74 | |
| Consanguinity/ inbreeding | | | | |
| % consanguineous/total marriages | 5.0 ± 0.72 (0.86 − 14.22) | 14.8 ± 1.44 (2.85 − 41.74) | 10.6 ± 1.06 (0.86 − 41.74) | 6.04*** |
| % first cousins/ total marriages | 1.2 ± 0.18 (0.21 − 5.08) | 3.0 ± 0.34 (0.0 − 9.57) | 2.2 ± 0.23 (0 − 9.57) | 4.83*** |
| % second cousins/total marriages | 2.0 ± 0.33 (0.21 − 7.41) | 6.6 ± 0.76 (0.81 − 18.38) | 4.6 ± 0.53 (0.21 − 18.38) | 5.53*** |
| Average inbreeding coefficient, α‰ | 1.4 ± 0.20 (0.23 − 4.90) | 3.8 ± 0.38 (0.64 − 11.45) | 2.8 ± 0.28 (0.23 − 11.45) | 5.59*** |
| No. of parishes ($n ≥ 8$) | 28 | 37 | 65 | |

Note. All the data were averaged over the period 1851–1950. Range of values in parentheses, standard errors computed as standard deviations of 1,000 bootstrapped samples.
***Significantly different from zero at $P = 0.001$.

per. The number of individuals in each sample is taken as equal to the number tested times the coefficient of redundancy (c.r.) given in the third column of table 5.7.

The first hypothesis worth testing is the possible existence of genetic heterogeneity among the major sections of the valley. This could be due to natural selection in different environments or to historical stratification of migrations. It must have happened more than once that new people coming from elsewhere have settled in the gen-

TABLE 5.6. Phenotype frequencies of the individuals tested in all 74 parishes of the Parma valley

| Parish | Number tested | ABO | | | | | | MN | | | Rh | | | | | | | | |
|---|---|---|---|---|---|---|---|---|---|---|---|---|---|---|---|---|---|---|---|
| | | 0 | $A_1$ | $A_2$ | B | $A_1B$ | $A_2B$ | M | MN | N | rr | $R_1r$ | $R_2r$ | $R_0r$ | $R'r$ | $R_1R_1$ | $R_1R_2$ | $R_2R_2$ | Others |
| Lemignano | 20 | 12 | 6 | 1 | 1 | 0 | 0 | 8 | 9 | 3 | 1 | 8 | 1 | 1 | 0 | 6 | 3 | 0 | 0 |
| Alberi | 31 | 16 | 9 | 1 | 2 | 3 | 0 | 5 | 20 | 6 | 2 | 16 | 1 | 0 | 0 | 8 | 4 | 0 | 0 |
| Carignano | 28 | 19 | 6 | 2 | 1 | 0 | 0 | 9 | 12 | 7 | 2 | 8 | 1 | 0 | 0 | 10 | 7 | 0 | 0 |
| Corcagnano | 42 | 14 | 17 | 4 | 4 | 3 | 0 | 14 | 21 | 7 | 7 | 13 | 3 | 0 | 0 | 15 | 2 | 0 | 2 |
| Gaione | 37 | 15 | 12 | 3 | 5 | 1 | 1 | 16 | 18 | 3 | 13 | 7 | 6 | 0 | 0 | 6 | 4 | 1 | 0 |
| Marore | 24 | 12 | 6 | 3 | 3 | 0 | 0 | 8 | 7 | 9 | 4 | 10 | 2 | 1 | 0 | 4 | 2 | 0 | 1 |
| Porporano | 32 | 14 | 11 | 3 | 1 | 3 | 0 | 13 | 13 | 6 | 5 | 11 | 6 | 0 | 0 | 7 | 3 | 0 | 0 |
| San Ruffino | 33 | 18 | 6 | 0 | 8 | 1 | 0 | 16 | 12 | 5 | 4 | 12 | 1 | 3 | 0 | 8 | 5 | 0 | 0 |
| Vigatto | 21 | 9 | 7 | 2 | 0 | 1 | 2 | 5 | 9 | 7 | 3 | 9 | 2 | 1 | 1 | 3 | 0 | 0 | 2 |
| Bas. Goiano | 47 | 27 | 13 | 1 | 3 | 2 | 1 | 15 | 22 | 10 | 4 | 11 | 6 | 2 | 1 | 14 | 8 | 1 | 0 |
| Bas. Nova | 101 | 43 | 28 | 11 | 11 | 5 | 3 | 27 | 42 | 32 | 15 | 37 | 13 | 2 | 0 | 16 | 18 | 0 | 0 |
| Bannone | 39 | 9 | 19 | 4 | 6 | 1 | 0 | 12 | 19 | 8 | 1 | 17 | 2 | 2 | 0 | 11 | 6 | 0 | 0 |
| Mamiano | 32 | 15 | 10 | 3 | 2 | 2 | 0 | 10 | 10 | 12 | 3 | 8 | 3 | 1 | 0 | 10 | 7 | 0 | 0 |
| Arola | 31 | 14 | 7 | 4 | 4 | 2 | 0 | 5 | 20 | 6 | 3 | 14 | 1 | 0 | 0 | 8 | 5 | 0 | 0 |
| Casatico | 43 | 24 | 6 | 4 | 7 | 2 | 0 | 10 | 24 | 9 | 5 | 22 | 6 | 0 | 0 | 5 | 4 | 1 | 0 |
| Castrignano | 48 | 21 | 18 | 4 | 4 | 0 | 1 | 13 | 18 | 17 | 7 | 20 | 3 | 0 | 1 | 9 | 8 | 0 | 0 |
| Cozzano | 39 | 15 | 18 | 3 | 3 | 0 | 0 | 6 | 25 | 8 | 9 | 17 | 3 | 0 | 0 | 4 | 5 | 1 | 0 |
| Mattaleto | 23 | 11 | 3 | 2 | 5 | 1 | 1 | 4 | 12 | 7 | 3 | 6 | 3 | 0 | 0 | 9 | 2 | 0 | 0 |
| Quinzano | 23 | 9 | 8 | 3 | 2 | 0 | 1 | 3 | 14 | 6 | 4 | 10 | 0 | 0 | 1 | 5 | 3 | 0 | 0 |
| Riano | 45 | 17 | 20 | 1 | 2 | 2 | 3 | 14 | 21 | 10 | 10 | 21 | 4 | 1 | 0 | 5 | 4 | 0 | 0 |
| Strognano | 24 | 11 | 4 | 4 | 5 | 0 | 0 | 9 | 12 | 3 | 5 | 16 | 0 | 0 | 0 | 1 | 0 | 0 | 0 |
| Tordenaso | 28 | 11 | 10 | 4 | 3 | 0 | 0 | 9 | 12 | 7 | 4 | 10 | 4 | 1 | 0 | 6 | 4 | 0 | 0 |
| Torrechiara | 25 | 7 | 11 | 1 | 5 | 1 | 0 | 11 | 11 | 3 | 4 | 6 | 5 | 2 | 0 | 2 | 5 | 0 | 1 |
| Cavana | 34 | 12 | 14 | 5 | 1 | 0 | 2 | 8 | 17 | 9 | 6 | 13 | 3 | 0 | 0 | 8 | 4 | 0 | 0 |

| | | | | | | | | | | | | | | | | | | |
|---|---|---|---|---|---|---|---|---|---|---|---|---|---|---|---|---|---|---|
| Faviano | 35 | 10 | 14 | 6 | 4 | 1 | 0 | 13 | 16 | 6 | 1 | 16 | 4 | 1 | 1 | 3 | 8 | 1 | 0 |
| Lesignano | 51 | 24 | 19 | 3 | 4 | 1 | 0 | 12 | 25 | 14 | 9 | 13 | 10 | 0 | 0 | 12 | 5 | 2 | 0 |
| Mulazzano | 39 | 18 | 11 | 5 | 3 | 1 | 1 | 17 | 12 | 10 | 6 | 14 | 5 | 2 | 0 | 8 | 4 | 0 | 0 |
| Rivalta | 53 | 19 | 21 | 3 | 8 | 2 | 0 | 19 | 26 | 8 | 4 | 20 | 8 | 1 | 2 | 9 | 8 | 1 | 0 |
| S.M. del Piano | 50 | 23 | 17 | 5 | 3 | 1 | 1 | 18 | 18 | 14 | 11 | 14 | 8 | 2 | 0 | 8 | 7 | 0 | 0 |
| Stadirano | 24 | 7 | 10 | 1 | 5 | 1 | 0 | 9 | 13 | 2 | 3 | 8 | 5 | 0 | 1 | 5 | 1 | 1 | 0 |
| Calestano | 45 | 16 | 20 | 3 | 4 | 2 | 0 | 9 | 25 | 11 | 8 | 13 | 11 | 0 | 0 | 3 | 8 | 2 | 0 |
| Antreola | 22 | 10 | 7 | 2 | 0 | 2 | 1 | 5 | 12 | 5 | 3 | 5 | 2 | 0 | 0 | 10 | 2 | 0 | 0 |
| Lupazzano | 41 | 22 | 12 | 4 | 3 | 0 | 0 | 16 | 17 | 8 | 6 | 15 | 6 | 1 | 1 | 12 | 0 | 0 | 0 |
| Mozzano | 29 | 14 | 8 | 3 | 4 | 0 | 0 | 9 | 11 | 9 | 4 | 10 | 1 | 0 | 0 | 7 | 6 | 1 | 0 |
| Neviano Ard. | 56 | 23 | 23 | 4 | 6 | 0 | 0 | 12 | 30 | 14 | 3 | 22 | 7 | 1 | 0 | 14 | 7 | 2 | 0 |
| Orzale | 59 | 33 | 13 | 7 | 5 | 0 | 1 | 11 | 33 | 15 | 7 | 36 | 5 | 1 | 1 | 5 | 2 | 2 | 0 |
| Urzano | 24 | 11 | 10 | 2 | 1 | 0 | 0 | 10 | 9 | 5 | 2 | 11 | 6 | 1 | 0 | 3 | 1 | 0 | 0 |
| Anzolla | 43 | 9 | 22 | 3 | 5 | 4 | 0 | 8 | 19 | 16 | 5 | 20 | 3 | 0 | 0 | 7 | 6 | 2 | 0 |
| Carobbio | 44 | 21 | 13 | 4 | 6 | 0 | 0 | 12 | 20 | 12 | 3 | 20 | 3 | 0 | 0 | 11 | 6 | 1 | 0 |
| Cereto | 43 | 18 | 14 | 5 | 4 | 2 | 0 | 19 | 17 | 7 | 6 | 12 | 8 | 2 | 0 | 7 | 8 | 0 | 0 |
| Madurera | 46 | 11 | 11 | 3 | 13 | 8 | 0 | 17 | 20 | 9 | 11 | 17 | 8 | 1 | 0 | 7 | 2 | 0 | 0 |
| Moragnano | 43 | 14 | 20 | 3 | 6 | 0 | 0 | 10 | 21 | 12 | 10 | 11 | 10 | 0 | 0 | 9 | 3 | 0 | 0 |
| Musiara Inf. | 53 | 22 | 21 | 3 | 3 | 3 | 1 | 20 | 29 | 4 | 8 | 20 | 9 | 1 | 0 | 3 | 12 | 0 | 0 |
| Reno | 29 | 13 | 9 | 3 | 3 | 1 | 0 | 7 | 15 | 7 | 5 | 9 | 3 | 1 | 0 | 7 | 4 | 0 | 0 |
| Tizzano | 25 | 12 | 7 | 0 | 6 | 0 | 0 | 4 | 16 | 5 | 5 | 4 | 1 | 2 | 1 | 7 | 4 | 0 | 1 |
| Ballone | 24 | 9 | 12 | 1 | 1 | 1 | 0 | 14 | 7 | 3 | 3 | 11 | 1 | 0 | 0 | 8 | 1 | 0 | 0 |
| Beduzzo | 55 | 26 | 16 | 4 | 7 | 2 | 0 | 18 | 22 | 15 | 10 | 14 | 9 | 0 | 1 | 12 | 9 | 0 | 0 |
| Bosco | 50 | 34 | 4 | 9 | 2 | 1 | 0 | 31 | 11 | 8 | 5 | 15 | 3 | 0 | 0 | 15 | 7 | 0 | 5 |
| Corniglio | 40 | 14 | 15 | 4 | 5 | 2 | 0 | 16 | 18 | 6 | 5 | 13 | 8 | 3 | 0 | 4 | 6 | 0 | 1 |
| Grammatica | 38 | 10 | 18 | 7 | 2 | 1 | 0 | 10 | 23 | 5 | 4 | 17 | 3 | 1 | 0 | 8 | 4 | 1 | 0 |
| Marra | 59 | 27 | 25 | 4 | 3 | 0 | 0 | 24 | 21 | 14 | 10 | 24 | 10 | 1 | 1 | 6 | 6 | 1 | 0 |
| Mossale | 34 | 11 | 14 | 2 | 3 | 3 | 1 | 8 | 17 | 9 | 4 | 22 | 1 | 2 | 0 | 2 | 1 | 0 | 2 |

TABLE 5.6. Continued

| Parish | Number tested | ABO | | | | | | MN | | | Rh | | | | | | | | |
|---|---|---|---|---|---|---|---|---|---|---|---|---|---|---|---|---|---|---|---|
| | | O | $A_1$ | $A_2$ | B | $A_1B$ | $A_2B$ | M | MN | N | rr | $R_1r$ | $R_2r$ | $R_0r$ | $R'r$ | $R_1R_1$ | $R_1R_2$ | $R_2R_2$ | Others |
| Petrignacola | 21 | 7 | 9 | 3 | 2 | 0 | 0 | 6 | 11 | 4 | 3 | 9 | 2 | 0 | 0 | 0 | 6 | 1 | 0 |
| Signatico | 30 | 12 | 11 | 2 | 5 | 0 | 0 | 7 | 16 | 7 | 0 | 10 | 3 | 0 | 0 | 10 | 7 | 0 | 0 |
| Vestana | 33 | 14 | 14 | 1 | 2 | 2 | 0 | 18 | 11 | 4 | 5 | 14 | 5 | 0 | 0 | 5 | 3 | 1 | 0 |
| Villula | 27 | 14 | 8 | 4 | 1 | 0 | 0 | 8 | 14 | 5 | 4 | 11 | 1 | 0 | 0 | 7 | 3 | 1 | 0 |
| Nirone | 50 | 30 | 12 | 6 | 2 | 0 | 0 | 14 | 25 | 11 | 15 | 12 | 5 | 2 | 1 | 6 | 7 | 2 | 0 |
| Palanzano | 35 | 16 | 14 | 0 | 1 | 0 | 4 | 12 | 15 | 8 | 5 | 13 | 2 | 2 | 0 | 8 | 4 | 1 | 0 |
| Pratopiano | 26 | 6 | 6 | 3 | 9 | 1 | 1 | 8 | 17 | 1 | 1 | 6 | 5 | 0 | 0 | 5 | 9 | 0 | 0 |
| Ranzano | 50 | 29 | 12 | 2 | 7 | 0 | 0 | 14 | 20 | 16 | 9 | 22 | 5 | 0 | 1 | 7 | 4 | 0 | 2 |
| Trevignano | 54 | 21 | 13 | 13 | 6 | 0 | 1 | 18 | 24 | 12 | 4 | 12 | 2 | 3 | 1 | 17 | 9 | 1 | 5 |
| Vaestano | 16 | 10 | 1 | 1 | 4 | 0 | 0 | 5 | 6 | 5 | 3 | 5 | 2 | 1 | 0 | 3 | 1 | 0 | 0 |
| Vairo | 21 | 12 | 4 | 0 | 4 | 1 | 0 | 6 | 13 | 2 | 2 | 10 | 4 | 0 | 0 | 4 | 1 | 0 | 0 |
| Valcieca | 35 | 11 | 10 | 5 | 5 | 3 | 1 | 14 | 14 | 7 | 6 | 22 | 1 | 3 | 0 | 3 | 0 | 0 | 0 |
| Casarola | 75 | 26 | 32 | 5 | 7 | 2 | 3 | 10 | 33 | 32 | 18 | 32 | 2 | 1 | 2 | 9 | 10 | 1 | 0 |
| Ceda | 21 | 12 | 8 | 1 | 0 | 0 | 0 | 4 | 12 | 5 | 2 | 13 | 0 | 0 | 0 | 3 | 3 | 0 | 0 |
| Lugagnano | 56 | 21 | 17 | 8 | 4 | 3 | 3 | 27 | 17 | 12 | 9 | 21 | 1 | 0 | 1 | 19 | 4 | 1 | 0 |
| Monchio | 45 | 16 | 14 | 7 | 3 | 4 | 1 | 9 | 29 | 7 | 5 | 23 | 2 | 0 | 1 | 7 | 6 | 1 | 0 |
| Pianadetto | 25 | 9 | 12 | 1 | 1 | 1 | 1 | 11 | 10 | 4 | 4 | 9 | 0 | 0 | 0 | 7 | 5 | 0 | 0 |
| Riana | 41 | 22 | 13 | 1 | 3 | 2 | 0 | 10 | 19 | 12 | 0 | 10 | 2 | 0 | 0 | 16 | 11 | 2 | 0 |
| Rigoso | 30 | 10 | 12 | 6 | 1 | 1 | 0 | 4 | 25 | 1 | 3 | 16 | 1 | 0 | 0 | 5 | 5 | 0 | 0 |
| Rimagna | 42 | 10 | 4 | 3 | 13 | 11 | 1 | 14 | 22 | 6 | 6 | 14 | 6 | 2 | 0 | 6 | 7 | 1 | 0 |
| Trefiumi | 43 | 20 | 15 | 3 | 5 | 0 | 0 | 26 | 13 | 4 | 8 | 21 | 1 | 0 | 0 | 8 | 3 | 2 | 0 |
| Valditacca | 35 | 18 | 11 | 4 | 2 | 0 | 0 | 9 | 16 | 10 | 8 | 15 | 1 | 0 | 0 | 6 | 1 | 4 | 0 |

eral region. Newcomers provided with better weapons than the former settlers were at special advantage over them, and would probably occupy the most attractive areas. Among prehistorical events of the kind must have been the first arrival of Neolithic farmers, most probably from the south and east. Mesolithic and Neolithic farmers were probably interested in different types of environments, since only a few areas, more likely in the plains and lower hills, were attractive for the Neolithic type of farming, while midaltitude forests and higher mountain areas may have been good hunting grounds for Mesolithics.

Major successive waves of new settlers after the Neolithic must have been those of the Bronze Age, mostly in the second millennium B.C. The "Ice Age man" or Similhaun man, nicknamed Oetzi, of the third millennium B.C., was a typical representative of the early Bronze Age of northern Italy. His mtDNA showed him to be genetically similar to modern people of the general area in which he was found, at the boundary between Austria and Italy, and also further north in the German plains (De Benedetto et al. 2000). He was probably living in a lower part of the country, on one or the other side of the Italian–Austrian border, and was perhaps trying to cross the Alps at the time of his death to escape from some personal danger. New settlers of the Iron Age must have started to arrive at the end of the second millennium B.C. At the beginning of the first millennium B.C. the Hallstatt Iron Age culture appears first in Austria. A later iron center is at La Tène in Switzerland. These settlers mark the beginning of the Celtic culture, which in the second half of the millennium extended to a large fraction of Central Europe and further, specifically to northern Italy and northern Spain. Rome was sacked by Gauls in 390 B.C.

It was therefore interesting to test the possibility of a genetic stratification indicating that different people live today in the various altitudes of the Parma valley. This was done by comparing the gene frequencies in P + H versus M, shown in table 5.8 by $2 \times n$ chi squares, using as observed numbers the gene frequencies multiplied by the total numbers of individuals times the weighted average c.r. coefficient. It is clear that there is no major genetic heterogeneity in the valley; if there was initially, it was diluted by later migration. It should be noted that the number of genes tested, the most important and informative ones available at the time, is very limited. Later

TABLE 5.7. Gene frequencies × 1000 of the individuals tested in the 74 parishes of the Parma valley

| Parish | Number tested | c.r. | ABO | | | | MN | | | Rh | | | | |
|---|---|---|---|---|---|---|---|---|---|---|---|---|---|---|
| | | | $A_1$ | $A_2$ | $B$ | $O$ | $M$ | $N$ | $r$ | $R_1$ | $R_2$ | $R_0$ | $R'$ | Other |
| Lemignano | 20 | 0.91 | 170 | 31 | 25 | 774 | 625 | 375 | 300 | 575 | 100 | 25 | 0 | 0 |
| Alberi | 31 | 0.94 | 219 | 22 | 83 | 677 | 484 | 516 | 339 | 581 | 81 | 0 | 0 | 0 |
| Carignano | 28 | 0.98 | 119 | 41 | 18 | 821 | 536 | 464 | 232 | 625 | 143 | 0 | 0 | 0 |
| Corcagnano | 42 | 0.90 | 306 | 73 | 87 | 534 | 583 | 417 | 381 | 536 | 60 | 0 | 0 | 24 |
| Gaione | 37 | 0.84 | 215 | 72 | 100 | 613 | 676 | 324 | 527 | 311 | 162 | 0 | 0 | 0 |
| Marore | 24 | 0.96 | 147 | 76 | 65 | 711 | 479 | 521 | 438 | 417 | 83 | 21 | 42 | 0 |
| Porporano | 32 | 0.97 | 273 | 71 | 63 | 593 | 609 | 391 | 422 | 438 | 141 | 0 | 0 | 0 |
| San Ruffino | 33 | 0.88 | 112 | 00 | 147 | 740 | 667 | 333 | 364 | 500 | 91 | 45 | 0 | 0 |
| Vigatto | 21 | 0.90 | 218 | 69 | 58 | 655 | 452 | 548 | 500 | 357 | 48 | 24 | 24 | 48 |
| Bas. Goiano | 47 | 0.94 | 180 | 26 | 65 | 728 | 553 | 447 | 298 | 500 | 170 | 21 | 11 | 0 |
| Bas. Nova | 101 | 0.97 | 203 | 93 | 98 | 606 | 475 | 525 | 406 | 431 | 154 | 10 | 0 | 0 |
| Bannone | 39 | 0.89 | 340 | 75 | 97 | 488 | 551 | 449 | 295 | 577 | 103 | 26 | 0 | 0 |
| Mamiano | 32 | 0.92 | 226 | 64 | 64 | 646 | 469 | 531 | 281 | 547 | 156 | 16 | 0 | 0 |
| Arola | 31 | 0.50 | 173 | 83 | 101 | 642 | 484 | 516 | 339 | 564 | 97 | 0 | 0 | 0 |
| Casatico | 43 | 0.92 | 103 | 54 | 110 | 733 | 512 | 488 | 442 | 419 | 140 | 0 | 0 | 0 |
| Castrignano | 48 | 0.75 | 232 | 70 | 54 | 644 | 458 | 542 | 396 | 479 | 115 | 0 | 10 | 0 |
| Cozzano | 39 | 0.96 | 290 | 55 | 40 | 616 | 474 | 526 | 487 | 385 | 128 | 0 | 0 | 0 |
| Mattaleto | 23 | 0.90 | 100 | 75 | 164 | 661 | 435 | 565 | 326 | 565 | 109 | 0 | 0 | 0 |
| Quinzano | 23 | 0.98 | 230 | 121 | 68 | 581 | 435 | 565 | 413 | 500 | 65 | 0 | 22 | 0 |

| | | | | | | | | | | | | | |
|---|---|---|---|---|---|---|---|---|---|---|---|---|---|
| Riano | 45 | 0.85 | 325 | 73 | 79 | 522 | 544 | 456 | 511 | 389 | 89 | 11 | 0 | 0 |
| Strognano | 24 | 1.00 | 098 | 98 | 112 | 693 | 625 | 375 | 563 | 375 | 42 | 21 | 0 | 0 |
| Tordenaso | 28 | 0.85 | 226 | 97 | 56 | 622 | 536 | 464 | 393 | 464 | 143 | 0 | 0 | 0 |
| Torrechiara | 25 | 0.98 | 294 | 28 | 131 | 548 | 660 | 340 | 440 | 300 | 200 | 40 | 0 | 20 |
| Cavana | 34 | 0.85 | 324 | 179 | 45 | 452 | 485 | 515 | 412 | 485 | 103 | 0 | 0 | 0 |
| Faviano | 35 | 0.96 | 291 | 127 | 75 | 506 | 600 | 400 | 343 | 429 | 200 | 14 | 14 | 0 |
| Lesignano | 51 | 0.97 | 232 | 39 | 50 | 678 | 480 | 520 | 402 | 412 | 186 | 0 | 0 | 0 |
| Mulazzano | 39 | 0.96 | 192 | 102 | 66 | 640 | 590 | 410 | 423 | 436 | 115 | 26 | 0 | 0 |
| Rivalta | 53 | 0.80 | 262 | 39 | 100 | 599 | 604 | 396 | 368 | 434 | 170 | 9 | 19 | 0 |
| S.M. del Piano | 50 | 0.96 | 224 | 82 | 51 | 643 | 540 | 460 | 460 | 370 | 150 | 20 | 0 | 0 |
| Stadirano | 24 | 0.96 | 278 | 28 | 136 | 558 | 646 | 354 | 417 | 396 | 167 | 0 | 21 | 0 |
| Calestano | 45 | 0.90 | 306 | 50 | 69 | 575 | 478 | 522 | 444 | 300 | 256 | 0 | 0 | 0 |
| Antreola | 22 | 0.90 | 205 | 64 | 57 | 674 | 500 | 500 | 296 | 614 | 91 | 0 | 0 | 0 |
| Lupazzano | 41 | 0.88 | 172 | 61 | 38 | 730 | 598 | 402 | 427 | 476 | 73 | 12 | 12 | 0 |
| Mozzano | 29 | 0.91 | 162 | 64 | 72 | 701 | 500 | 500 | 328 | 517 | 155 | 0 | 0 | 0 |
| Neviano Ard. | 56 | 0.86 | 250 | 48 | 56 | 646 | 482 | 518 | 321 | 509 | 161 | 9 | 0 | 0 |
| Orzale | 59 | 0.90 | 130 | 81 | 52 | 737 | 466 | 534 | 483 | 407 | 93 | 9 | 9 | 0 |
| Urzano | 24 | 0.90 | 256 | 58 | 21 | 665 | 604 | 396 | 458 | 375 | 146 | 21 | 0 | 0 |
| Anzolla | 43 | 0.74 | 409 | 60 | 112 | 419 | 407 | 593 | 384 | 465 | 151 | 0 | 0 | 0 |
| Carobbio | 44 | 0.89 | 173 | 56 | 72 | 699 | 500 | 500 | 329 | 545 | 125 | 0 | 0 | 0 |
| Cereto | 43 | 0.93 | 230 | 80 | 72 | 617 | 640 | 361 | 395 | 395 | 186 | 23 | 0 | 0 |
| Madurera | 46 | 0.83 | 244 | 46 | 260 | 450 | 587 | 413 | 522 | 359 | 109 | 11 | 0 | 0 |
| Moragnano | 43 | 0.91 | 291 | 49 | 74 | 586 | 477 | 523 | 477 | 372 | 151 | 0 | 0 | 0 |
| Musiara Inf. | 53 | 0.83 | 281 | 56 | 68 | 595 | 651 | 349 | 434 | 359 | 198 | 9 | 0 | 0 |

TABLE 5.7. Continued

| Parish | Number tested | c.r. | ABO | | | | MN | | Rh | | | | | |
|---|---|---|---|---|---|---|---|---|---|---|---|---|---|---|
| | | | $A_1$ | $A_2$ | $B$ | $O$ | $M$ | $N$ | $r$ | $R_1$ | $R_2$ | $R_0$ | $R'$ | Other |
| Reno | 29 | 0.98 | 208 | 68 | 72 | 652 | 500 | 500 | 397 | 465 | 121 | 17 | 0 | 0 |
| Tizzano | 25 | 0.94 | 154 | 0 | 130 | 716 | 480 | 520 | 360 | 460 | 100 | 40 | 20 | 20 |
| Ballone | 24 | 0.94 | 338 | 33 | 42 | 587 | 729 | 271 | 375 | 583 | 42 | 0 | 0 | 0 |
| Beduzzo | 55 | 0.97 | 190 | 46 | 86 | 678 | 527 | 473 | 400 | 427 | 164 | 0 | 9 | 0 |
| Bosco | 50 | 0.92 | 57 | 101 | 30 | 811 | 730 | 270 | 280 | 560 | 100 | 0 | 0 | 60 |
| Corniglio | 40 | 0.94 | 266 | 71 | 92 | 570 | 625 | 375 | 425 | 338 | 175 | 38 | 25 | 0 |
| Grammatica | 38 | 0.82 | 374 | 158 | 41 | 427 | 566 | 434 | 382 | 487 | 118 | 13 | 0 | 0 |
| Marra | 59 | 0.82 | 258 | 47 | 26 | 670 | 585 | 415 | 475 | 356 | 153 | 9 | 9 | 0 |
| Mossale | 34 | 0.99 | 328 | 73 | 107 | 491 | 485 | 515 | 485 | 426 | 29 | 29 | 0 | 29 |
| Petrignacola | 21 | 0.95 | 283 | 103 | 50 | 564 | 548 | 452 | 405 | 357 | 238 | 0 | 0 | 0 |
| Signatico | 30 | 0.92 | 218 | 43 | 89 | 650 | 500 | 500 | 217 | 617 | 167 | 0 | 0 | 0 |
| Vestana | 33 | 0.95 | 289 | 22 | 62 | 627 | 712 | 288 | 439 | 409 | 152 | 0 | 0 | 0 |
| Villula | 27 | 0.98 | 182 | 96 | 19 | 703 | 556 | 444 | 370 | 518 | 111 | 0 | 0 | 0 |
| Nirone | 50 | 0.80 | 140 | 73 | 20 | 767 | 530 | 470 | 500 | 310 | 160 | 20 | 10 | 0 |
| Palanzano | 35 | 0.93 | 250 | 0 | 74 | 676 | 557 | 443 | 386 | 471 | 114 | 29 | 0 | 0 |
| Pratopiano | 26 | 0.81 | 164 | 95 | 246 | 495 | 635 | 365 | 250 | 481 | 269 | 0 | 0 | 0 |
| Ranzano | 50 | 0.91 | 132 | 23 | 73 | 771 | 480 | 520 | 470 | 410 | 90 | 0 | 10 | 20 |
| Trevignano | 54 | 0.91 | 160 | 170 | 68 | 603 | 556 | 444 | 241 | 546 | 120 | 28 | 9 | 56 |

| | | c.r. | | | | | | | | | | | | |
|---|---|---|---|---|---|---|---|---|---|---|---|---|---|---|
| Vaestano | 16 | 0.94 | 33 | 33 | 135 | 799 | 500 | 500 | 438 | 375 | 156 | 31 | 0 | 0 |
| Vairo | 21 | 0.88 | 126 | 0 | 126 | 747 | 595 | 405 | 429 | 452 | 119 | 0 | 0 | 0 |
| Valcieca | 35 | 0.80 | 247 | 126 | 137 | 489 | 600 | 400 | 543 | 400 | 14 | 43 | 0 | 0 |
| Casarola | 75 | 0.84 | 297 | 80 | 83 | 540 | 353 | 647 | 487 | 400 | 93 | 7 | 13 | 0 |
| Ceda | 21 | 0.93 | 222 | 31 | 0 | 747 | 476 | 524 | 405 | 524 | 71 | 0 | 0 | 0 |
| Lugagnano | 56 | 0.95 | 251 | 150 | 92 | 506 | 634 | 366 | 366 | 563 | 63 | 0 | 9 | 0 |
| Monchio | 45 | 0.92 | 282 | 146 | 92 | 480 | 522 | 478 | 400 | 478 | 111 | 0 | 11 | 0 |
| Pianadetto | 25 | 0.94 | 345 | 67 | 61 | 528 | 640 | 360 | 340 | 560 | 100 | 0 | 0 | 0 |
| Riana | 41 | 0.93 | 206 | 16 | 63 | 716 | 476 | 524 | 146 | 646 | 207 | 0 | 0 | 0 |
| Rigoso | 30 | 0.95 | 321 | 169 | 34 | 476 | 550 | 450 | 383 | 517 | 100 | 0 | 0 | 0 |
| Rimagna | 42 | 0.94 | 211 | 73 | 347 | 369 | 595 | 405 | 405 | 393 | 179 | 24 | 0 | 0 |
| Trefiumi | 43 | 0.93 | 206 | 45 | 61 | 689 | 756 | 244 | 442 | 465 | 93 | 0 | 0 | 0 |
| Valditacca | 35 | 0.93 | 189 | 73 | 29 | 708 | 486 | 514 | 057 | 400 | 43 | 0 | 0 | 0 |

*Note.* c.r., correction for redundancy.

TABLE 5.8. Frequencies of alleles for ABO, MN, and Rh systems according to the altitude of the parishes

| | ABO | | | | MN | | | Rh | | | | | |
| | $A_1$ | $A_2$ | $B$ | $O$ | $M$ | $N$ | $r$ | $R_1$ | $R_2$ | Other | $2n$ | c.r. |
|---|---|---|---|---|---|---|---|---|---|---|---|---|
| Plains + Hills | 0.218 (0.011) | 0.067 (0.005) | 0.075 (0.006) | 0.639 (0.013) | 0.538 (0.012) | 0.462 (0.012) | 0.397 (0.013) | 0.459 (0.014) | 0.127 (0.008) | 0.017 (0.000) | 2,756 | 0.9018 |
| Mountains | 0.231 (0.014) | 0.069 (0.008) | 0.089 (0.011) | 0.611 (0.019) | 0.561 (0.015) | 0.439 (0.015) | 0.396 (0.014) | 0.456 (0.014) | 0.130 (0.009) | 0.018 (0.000) | 2,874 | 0.8937 |
| $\chi^2$ | 5.72 | | | | 2.67 | | | 0.19 | | | | |
| d.f. | 3 | | | | 1 | | | 3 | | | | |
| $P$ | <0.10 | | | | 0.10 | | | >0.50 | | | | |

Note. Values in parentheses are standard errors. c.r., correction for redundancy.

work could have used a much wider sample of genes and might have produced a different result.

A principal component analysis (PCA, see Cavalli-Sforza et al. [1994], chapter 1 section 13) of genetic variation in Italy, using a much greater number of genetic markers (Piazza et al. 1988), found a clear difference between people living in the western portion of the Ligurian Apennines, located north and west of Parma, and the rest of the Italian population. In this region of the Ligurian Apennines, belonging to the provinces of Genova, Alessandria, Pavia, and Piacenza, there is a fairly wide mountainous area, which is separated from the Parma valley by deep valleys and relatively low mountain passes. This would have been a perfect refuge for Ligurians, one of the most ancient Italian peoples, who prior to the arrival of Celts, occupied a wide region: all the western part of northern Italy and the Rhone valley in the southeastern part of France. It is possible that Ligurians in this area were less exposed to admixture with late-comers, such as the Etruscans and the Celts, followed by Romans who conquered most of northern Italy at the end of the third century B.C. There are extremely few written remnants of the Ligurian language, mostly in an Etruscan alphabet that was spread in northern Italy around the sixth century B.C. Some linguists classify it as pre-Indo-European; others consider it as Celtic, but this may be a result of later borrowings, as is likely to have happened in Spain for "celt-iberian," when Iberians, speaking an earlier pre-Indo-European language, mixed with later-arriving Celts.

A further problem is a comparison of the overall genetic variation existing in the P + H and the M subsets. This was done in two different ways, leading to results of different meaning. In both cases the $F_{ST}$ genetic distance was used. This is calculated as a variance of gene frequencies, standardized by division by the product of the mean gene frequency over the whole area investigated, and one minus the mean gene frequency. The exact formula employed is that given by Reynolds et al. (1983) (see Cavalli-Sforza et al. [1994], chapter 1 section 10). The existence of variation between villages within each of the two areas was first calculated directly by an $F_{ST}$ distance among all the villages within each of the two areas; the two measures of heterogeneity cumulating in the three blood group systems are given in table 5.9, first line. It is immediately clear that there

TABLE 5.9. Parma valley: genetic variation measured by $F_{ST}$ distance

| | Plains + Hills | Mountains | Total | P + H vs. M t-test |
|---|---|---|---|---|
| Average $F_{ST}$ distance among the parishes of the regions | 0.0076 ± 0.0021 | 0.0221 ± 0.0040 | 0.0150 ± 0.0025 | 3.12** |
| Average $F_{ST}$ distance from nearest neighbors | 0.0046 ± 0.0011 | 0.0109 ± 0.0025 | 0.0078 ± 0.0014 | 2.30* |
| Number of parishes | 37 | 37 | 74 | |

Note. Standard errors computed as standard deviations of 1000 bootstrapped samples.
*Significantly different from zero at P = 0.05.
**Significantly different from zero at P = 0.01.

is much greater variation among villages in the mountain area. The difference of $F_{ST}$ between P + H and M is clearly significant:

$$t = 3.12, \quad P < 0.001$$

Since P + H and M are originally variance values, it may be questioned if the use of $t$ is appropriate. But because they are obtained by averaging $F_{ST}$ from many alleles, they share the properties of means, which by the central limit theorem tend to be normally distributed as soon as they derive from the sum of many values.

The second line of table 5.9 shows the results of a method that is a new attempt at evaluating "local" drift. One calculates the difference between each village and the immediately adjacent ones, again in the form of an $F_{ST}$ distance. The quantity thus calculated for each village will be referred to as distance between a village and its nearest neighbors ($V - NN$) located within 5 km, averaged over all villages. These values have some internal autocorrelation, because the gene frequencies of each village are used repeatedly in the calculation. It was nevertheless considered interesting to try their use later for correlation studies.

The difference between P + H and M is significant also with this new measure, though at a lower level:

$$t = 2.30, \quad P < 0.05$$

It is clear that there is a much higher drift in the mountains. We proceed to examine the same question with an entirely different set of data, surnames, which are more informative because they use a greater number of "alleles."


## 5.5 SURNAMES AND GENETIC DRIFT


Data on gene frequencies obtainable when the Parma valley research was started were limited by the available gene systems, which then were only the major blood groups. It later became clear that an easily available marker, surnames, could be used for the same purpose. This marker is limited to male transmission, and recent studies (Seielstad et al. 1998) have shown that there is on average a marked difference in the migration of the two sexes. As mentioned earlier, the difference is clearly revealed by comparing two types of genetic data: that of mitochondrial DNA, which is transmitted only by females, and that of Y-chromosome markers, which are transmitted only by males.

The surnames employed here come from the list of consanguineous marriages that was obtained from the archives of consanguineous marriage dispensations collected in the Parma valley for the period 1851–1950. The use of surnames of consanguineous marriages may introduce a slight bias with results that could be obtained with surnames of the general population, but this bias probably will not affect the evolutionary factors in which we are mostly interested. We prefer to use the surnames of females, as we did in earlier studies using surnames as markers of genetic variation, because they reflect more closely the residence of the spouse. Female surnames are also transmitted by males, but when using parish books of marriages as the source of surnames, as we usually did, those of females show less geographic variation because marriages are recorded in the wives' parish. In using the surnames of females we are likely to come a little closer to an average of the two sexes, but the estimates of migration that we obtain from surnames will be closer to those of male transmitted characters than to those of standard genetic markers.

For the analysis of drift based on the frequencies of surnames we use the same methods developed for gene frequencies, and therefore calculate the same $F_{ST}$ distances as in the last section. The first analysis made there was based on the variance of frequencies of alleles (here replaced by surnames) among all the parishes of the two subregions and the whole valley. Results of table 5.10 show higher significance than for blood group genes. The test for drift using $F_{ST}$ among all parishes is more powerful than the comparison of each parish with its nearest neighbors when using blood group genes, but it is more slightly powerful when using surnames (table 5.10).

Surnames allow us to calculate an alternative measure of drift. We have tested it in a variety of other circumstances and found it potentially very useful. We have here a chance to test it further. This measure can be called "abundance of surnames" and derives from a theory by R. A. Fisher (1943) on the estimation of species abundance. In our circumstances it is comparable to the quantity $Nm$, a basic estimate of population structure, where $N$ is the size of the deme (see chapter 10), the average effective population size, $N_e$, and $m$ is the

TABLE 5.10. Parma valley: surname variation measured by $F_{ST}$ distance

| | Plains + Hills | Mountains | Total | $P + H$ vs. M t-test |
|---|---|---|---|---|
| Average $F_{ST}$ distance among the parishes of the regions | $0.0577 \pm 0.0059$ | $0.1041 \pm 0.0121$ | $0.0897 \pm 0.0084$ | 3.45*** |
| Average $F_{ST}$ distance from nearest neighbors | $0.0287 \pm 0.0044$ | $0.0555 \pm 0.0054$ | $0.0442 \pm 0.0040$ | 3.84*** |
| Surname abundance | $28.5 \pm 4.06$ $(2.3 - 96.5)$ | $11.9 \pm 1.28$ $(2.9 - 34.6)$ | $18.7 \pm 2.09$ $(2.3 - 96.5)$ | 3.90*** |
| Number of parishes | 26 | 37 | 63 | |

*Note.* Range of values in parentheses; standard errors computed as standard deviations of 1,000 bootstrapped samples.
***Significantly different from zero at $P = 0.001$.

fraction of population exchanged every generation by migration between a deme and its neighbors. For the ways of estimating abundance of surnames, see Zei et al. (1983b) and chapter 10. The values
of surname abundance obtained from this material are also given in
table 5.10 third line, and the differences are highly significant.

It is important to note that surname abundance is in an inverse
relationship to $F_{ST}$ values (Zei et al. 1996). The expected values of $F_{ST}$
at equilibrium are $1/(1 + 4Nm)$ for standard genes and $1/(1 + Nm)$
for surnames. In fact, for P + H + M, with $Nm = 18.7$ from surname abundance, the expected $F_{ST}$ for blood group genes is 0.0132,
which is in good agreement with the observed value of $0.0150 \pm
0.0025$ for $F_{ST}$ among all parishes. For surnames, the expectation of
$F_{ST}$ from $Nm = 18.7$ is 0.0508, which is in less good (but still acceptable) agreement with that calculated directly, $0.0897 \pm 0.0084$.

$F_{ST}$ values obtained on widely different markers like surnames
and genes cannot be compared directly, and hence there is no surprise
if the $F_{ST}$ values of surnames and blood group genes are quite different. The ratio of $F_{ST}$ values of the two geographic areas, M and
P + H, however, is comparable:

|  | $F_{ST}$ M $/ F_{ST}$ (P + H) |
|---|---|
| $F_{ST}$ among all parishes |  |
| Genes | 2.91 |
| Surnames | 1.80 |
|  |  |
| $F_{ST}$ V − NN |  |
| Surnames | 2.37 |
| Genes | 1.93 |
|  |  |
| Surname abundance M $/ (P + H)$ | 2.39 |

The values for the same markers give reasonably close ratios. Drift in
mountains is from 2 to 3 times greater than in the plains + hills.

## 5.6 CORRELATIONS OF INBREEDING AND DRIFT

Inbreeding and drift are not exactly the same thing, but are very close
to each other. At equilibrium, one expects the average inbreeding
coefficient to be equal to $1/(1 + 4Nm)$, and this is the same expression valid for the genetic variance expressed by $F_{ST}$.

There are some limitations to this expectation. The equilibrium value of $F_{ST}$ under the opposite forces of migration and drift is reached fairly rapidly if migration is high, as is true of most human populations. We will see in the section dedicated to population simulations that this is valid in the present circumstances. But there are changes of migration and population size. Population sizes (table 5.1) increased by 45% in the mountain area, where drift is most easily observed, from 1861 to 1911, and decreased by 38% in the next 50 years. Changes from the earliest times for which parish books are available were less drastic. Migration also changed but it is difficult to have good estimates for different times.

The changes of inbreeding were more serious and consanguinity has been increasing during much of the nineteenth century, but has been constantly decreasing since 1915. Estimates of inbreeding are affected by a more serious source of error: recorded relationships are only the tip of the iceberg. It is true that the inbreeding coefficient of specific matings decreases rapidly as the number of generations separating two relatives increases, but the number of hidden relationships may increase considerably under those same circumstances. Thus, the average inbreeding tends to increase with the number of earlier generations included in the analysis, but it is difficult to make accurate estimates because the reconstruction of genealogies is usually limited to the very last generations. An example in which the contribution to the average inbreeding has been estimated separately for first + second and for third + fourth cousins (in the Hutterites of South Dakota) gave, respectively, 0.010 and 0.012, totaling 0.022 (Mange 1964). The average inbreeding also depends on population demography, and the Hutterites are an unusually fast-growing population.

The average inbreeding coefficient of the Parma valley per thousand is (see table 5.5):

P + H:          1.4 ± 0.20
M:              3.8 ± 0.38
P + H + M:      2.8 ± 0.28

The ratio between M and P + H is 2.71, which compares favorably with values of this ratio for the $F_{ST}$ obtained above from genes and surnames.

The average inbreeding coefficient, at equilibrium, should be the same as $F_{ST}$ from blood group genes, which is 0.015 for the whole area. It is, however, smaller in a ratio of 5.4:1. This big discrepancy is due to the effect of many sources of error. The most important one must be the incompleteness of the record of the more remote relationships, which even in the most favorable cases do not include fourth cousins. Can this explain the whole gap? It seems unlikely, since this would mean that only 18% of the existing consanguinity is recorded.

It is possible, and indeed likely, that part of the gap is due to avoidance of consanguineous marriages because of fear of giving birth to progeny likely to die early or be disabled, and/or because of the restrictions imposed by the Catholic Church. We suggested earlier, on the basis of a preliminary analysis (Cavalli-Sforza et al. 1966) that avoidance reduces by a factor of 2 the number of consanguineous marriages taking place. If this is correct, then the current consanguinity estimates record only 36% of the existing inbreeding, and 64% are lost mostly because they occur in earlier generations. This estimate could be modified by further considerations, as we see in the next chapter. It remains clear, however, that the average inbreeding is very seriously underestimated by consanguinity records, at least in the present case.

The correlations between the four basic measures among all the parishes we have studied are collected in table 5.11. They are all significantly different from zero, with that between inbreeding coefficient of a village and its genetic distance from neighbors (V − NN) being lowest (0.312). This is in part because the estimate of genetic distance with only three blood group systems is not sufficiently accurate. In fact, the same correlation calculated using surname distance, which is undoubtedly more accurate (because it is based on a much larger number of alleles), is definitely higher (0.537). Note that the sign of correlations between surname abundance and the other three variables is negative. This is expected because, as already mentioned, surname abundance is inversely related with all the other variables and is closest to $Nm$.

There are two major conclusions to be drawn from the investigations described in this chapter. In the mountains of the Parma valley the amount of genetic drift is at least two to three times greater than in the hills or plains, where village size is much larger. It remains to be seen if it corresponds to the amount of drift expected on the basis

TABLE 5.11. Correlations between estimates obtained from consanguinity, genetics, and two methods using surname data (Fisher surname abundance and $F_{ST}$ distance)

|  | Inbreeding coefficient $\alpha$ | Genetic $F_{ST}$ distance | Surname $F_{ST}$ distance |
|---|---|---|---|
| Genetic $F_{ST}$ distance | 0.312 (0.012) | | |
| Surname $F_{ST}$ distance | 0.537 (<0.001) | 0.422 (<0.001) | |
| Surname abundance | −0.575 (<0.001) | −0.336 (<0.007) | −0.836 (<0.001) |

Note. P values in parentheses.

of demographic knowledge. This is discussed in the next chapter. Another major conclusion is that the inbreeding coefficient estimated from consanguineous marriages is too low. The underestimate is by a factor of five. This is likely to be a consequence, in part, of ignoring remote consanguinity, in part of the partial avoidance of consanguineous marriages. The next chapter will help increase our understanding of the sources of this discrepancy.

# CHAPTER 6

# A Computer Simulation of the Upper Parma Valley Population

## 6.1 THE NEED FOR A POPULATION SIMULATION

The two major problems that we have tried to attack in this volume are the measurement of inbreeding in human populations, through a study of consanguinity, and that of random genetic drift. We have obtained some measurements and developed new methods that have been illustrated in the last three chapters, but we have not yet given a full answer to some major questions.

It soon became clear that a new approach could be very helpful, that of creating by computer an artificial population as similar as possible to the real one and estimating genetic quantities from it, which could then be compared to those of the real population. First runs were made in 1963. We realized that with the very limited computer memory available in the computer used at the time (Olivetti Elea) we could simulate at most a population of about 5000 individuals. We therefore chose for simulation the set of 22 villages in the highest portion of the Parma valley, whose inhabitants sum to a very similar figure.

The original program was written in Elea machine language by the Olivetti staff. With a small memory and three tapes the simulation was slow. The simulation program was later rewritten by us in Fortran for an IBM 7040. The first results of the simulation were communicated in 1966 to the Third International Congress of Human Genetics (Chicago, 1966) and published only very partially in two papers, and especially in Cavalli-Sforza and Zei (1967). The comparison with the real data was partially discussed in Cavalli-Sforza and Bodmer (1971, 1999) and has been completed only now.

In this chapter we summarize the organization of the simulation (section 6.2) in ways that would make it possible to reproduce it for this or other populations. We also compared expected drift with observed drift, and expected and observed consanguinities, expectations being always computed from the artificial populations. Comparison of observed and expected drift were also done with expectations calculated by the migration matrix method (Bodmer and Cavalli-Sforza 1968), especially in pages 433–474 of Cavalli-Sforza and Bodmer (1971, 1999). Some disagreement between observed and expected drift were noted. Imperfect matching of the populations being compared and the use of a relatively inefficient estimate of $F_{ST}$ from gene frequencies (called Wahlund's variance in Cavalli-Sforza and Bodmer [1971]) are mostly responsible for the disagreement. Expectations from the artificial population and migration matrix approach were similar, but the artificial population values allow the evaluation of many other potential errors. In the present treatment we have found sources of disagreement in the earlier analysis and made corrections to assure a better comparability of estimates.

## 6.2 STRUCTURE OF THE SIMULATION

The simulation included 22 villages with an expected total of 5024 inhabitants. Naturally, real numbers fluctuated at every cycle. Three matrices were defined: (1) married individuals with numerical code, identification number, sex, and other possible characters, (2) single males, and (3) single females. There was a code for social position but, as it had little effect, we will ignore it in what follows.

The first step was the creation of a simplified initial population, made of only two generations:

1. Parental pairs were generated first, and age and genes were assigned to them (only ABO, MN, RH) according to demographic tables and average gene frequency data. All ages were given in ten-year intervals thus, codes 0 = 0–9 years old,

1 = 10–19 years old, 2 = 20–29 years old, and so on. The ages of husbands and wives were initially the same and were given according to table 6.1. This and other tables relating to the artificial population values are given as they are used in the computer, that is, as the cumulative distributions, used directly with random numbers in the range 0–99, 0–999, and so on, depending on the number of digits used in the table. Thus, as an example, if the random number generating the first individual is 803, by looking at table 6.1 one assigns age 4 (=40–49 years) to the individual because 803 is >659 (the upper limit of the class corresponding to age 3) and =859 (the upper limit of the class corresponding to age 4). The gene frequencies used for generating genotypes were those of the real population according to table 6.2, and all individuals received genes at random so that there was no initial heterogeneity between villages.

2. Children were generated in numbers and ages depending on age of father (tables not given), and assigned ABO, MN, or RH genotype according to Mendelian rules, given the genotype of the parents. The numerical codes of the father and mother were included in the information of each child.

The time unit was of 10 years. This generated some approximation of demographic parameters, which are usually given statistically per year or every 5 years of age, but speeded up the process. At every cycle, three routines took place:

1. *Death and age subroutine.* For all individuals of all villages, survival as a function of age was ascertained by entering random

TABLE 6.1. Probability of marriage at given age for husband and wife

|  | Age | | | | |
|---|---|---|---|---|---|
|  | −19 | 20–29 | 30–39 | 40–49 | 50–59 |
|  | 1 | 2 | 3 | 4 | 5 |
| Cumulative probability × $10^3$ | 219 | 451 | 659 | 859 | 999 |

TABLE 6.2. Probability for a newborn to receive a specific allele of the ABO, MN, and RH systems

**ABO**

|  | $A_1$ 1 | $A_2$ 2 | $B$ 3 | $O$ 4 |
|---|---|---|---|---|
| Cumulative probability $\times 10^2$ | 19 | 26 | 34 | 99 |

**MN**

|  | $M$ 1 | $N$ 2 |
|---|---|---|
| Cumulative probability $\times 10^2$ | 54 | 99 |

**RH**

|  | $R_z$ 1 | $R_1$ 2 | $R'$ 3 | $R_2$ 4 | $R''$ 5 | $R_0$ 6 | $r$ 7 |
|---|---|---|---|---|---|---|---|
| Cumulative Probability $\times 10^3$ | 8 | 408 | 444 | 544 | 622 | 632 | 999 |

numbers into table 6.3. Age of survivors was increased by one. The places of deceased individuals were left vacant, and, if they were married, their widows/widowers were transferred into the respective singles matrices.

2. *Marriage subroutine.* All single men were tested for marriage, first with a woman from "abroad" (with probability 0.147, calculated as the average from the real villages). If the test was positive, the new woman was generated with age specified by the correlation table of age between spouses (table 6.4). If not, the test was repeated with

TABLE 6.3. Probability of death at a given age

|  | *Age* | | | | | |
|---|---|---|---|---|---|---|
|  | 0–9 0 | 10–19 1 | 20–29 2 | 30–39 3 | 40–49 4 | 50–59 5 |
| Cumulative probability $\times 10^2$ | 43 | 45 | 58 | 72 | 80 | 99 |

TABLE 6.4. Probability for a man of a given age to marry a woman of a
given age

|  | Age of wife | | | |
| --- | --- | --- | --- | --- |
| Age of husband | 20–29<br>2 | 30–39<br>3 | 40–49<br>4 | 50–59<br>5 |
| 20–29  2 | 799 | 999 | 999 | 999 |
| 30–39  3 | 889 | 969 | 999 | 999 |
| 40–49  4 | 319 | 679 | 959 | 999 |
| 50–59  5 | — | 329 | 499 | 999 |
| 60–69  6 | — | — | 249 | 999 |

a woman from a village chosen by the matrimonial migration matrix
(table 6.5), transformed in the cumulative form as in table 6.6. If a
single woman of the desired village was available, a further test was
made for desired age, according to table 6.4, and if more than one
woman of the desired age was available, she was chosen randomly.
For the marriage to be confirmed, room was first searched in the
future husband's village, or, if this was not available, in the wife's
village. If place was not available, the marriage was not confirmed
and a new marriage inside the valley was tried three times for the
same single man. If the marriage took place, spouses were transferred
to the married lists. The cycle was repeated for new individuals until
all single males went through the routine. Some remained unmarried
at the end.

   3. *Progeny generation subroutine.* The total number of children to
be born at every cycle, NB, was equated to the count of empty spaces
in single males and females matrices. The number of children per
married woman was in proportion 0.5, 0.3, and 0.2 per woman of
ages 2, 3, and 4, and the actual number was determined for each
woman by a Poisson distribution. If the total number of future children thus obtained, NC, was greater than the total number to be born,
NB, then some children of some mothers were eliminated randomly
until NC = NB. For each pair of spouses, genotypes were given to
the children born according to Mendel's rules and age 0. The age of
everybody else was increased by one at every cycle.

TABLE 6.5. Migration matrix

| Wife's birthplace | | 1 | 2 | 3 | 4 | 5 | 6 | 7 | 8 | 9 | 10 | 11 | 12 | 13 | 14 | 15 | 16 | 17 | 18 | 19 | 20 | 21 | 22 | Total |
|---|---|---|---|---|---|---|---|---|---|---|---|---|---|---|---|---|---|---|---|---|---|---|---|---|
| | | | | | | | | | | | | Husband's birthplace | | | | | | | | | | | | |
| Bosco | 1 | 1,044 | 0 | 20 | 3 | 0 | 45 | 2 | 18 | 0 | 1 | 1 | 0 | 0 | 2 | 1 | 0 | 0 | 0 | 0 | 0 | 0 | 1 | 1,138 |
| Casarola | 2 | 0 | 113 | 3 | 4 | 0 | 0 | 9 | 1 | 0 | 0 | 4 | 0 | 0 | 8 | 0 | 0 | 0 | 0 | 1 | 0 | 0 | 0 | 143 |
| Corniglio | 3 | 13 | 1 | 567 | 6 | 1 | 8 | 2 | 10 | 0 | 4 | 0 | 0 | 0 | 3 | 0 | 0 | 0 | 0 | 0 | 0 | 2 | 34 | 651 |
| Grammatica | 4 | 1 | 3 | 1 | 41 | 1 | 0 | 0 | 3 | 0 | 0 | 1 | 0 | 0 | 7 | 0 | 0 | 0 | 0 | 0 | 0 | 1 | 0 | 59 |
| Lugagnano | 5 | 0 | 3 | 0 | 0 | 201 | 0 | 22 | 0 | 4 | 13 | 1 | 0 | 3 | 0 | 2 | 4 | 2 | 0 | 1 | 7 | 2 | 0 | 265 |
| Marra | 6 | 32 | 1 | 9 | 1 | 1 | 107 | 0 | 5 | 0 | 0 | 0 | 0 | 0 | 0 | 0 | 0 | 0 | 0 | 1 | 0 | 0 | 3 | 160 |
| Monchio | 7 | 0 | 12 | 8 | 6 | 19 | 0 | 261 | 1 | 0 | 4 | 13 | 0 | 2 | 3 | 2 | 5 | 2 | 0 | 1 | 1 | 8 | 0 | 343 |
| Mossale | 8 | 21 | 2 | 16 | 0 | 0 | 7 | 0 | 162 | 0 | 0 | 0 | 0 | 0 | 2 | 0 | 0 | 1 | 0 | 0 | 0 | 0 | 1 | 218 |
| Nirone | 9 | 0 | 0 | 0 | 0 | 5 | 0 | 0 | 0 | 80 | 1 | 0 | 1 | 0 | 0 | 1 | 0 | 0 | 0 | 2 | 3 | 0 | 0 | 93 |
| Palanzano | 10 | 1 | 0 | 2 | 0 | 5 | 0 | 2 | 0 | 4 | 361 | 0 | 7 | 12 | 2 | 1 | 0 | 0 | 3 | 21 | 1 | 1 | 2 | 425 |
| Pianadetto | 11 | 1 | 4 | 3 | 0 | 1 | 0 | 30 | 0 | 0 | 0 | 170 | 0 | 0 | 4 | 7 | 6 | 6 | 1 | 0 | 0 | 40 | 0 | 273 |
| Pratopiano | 12 | 0 | 0 | 0 | 0 | 2 | 0 | 0 | 0 | 0 | 6 | 0 | 65 | 19 | 0 | 0 | 0 | 0 | 1 | 7 | 0 | 1 | 0 | 101 |
| Ranzano | 13 | 1 | 0 | 1 | 0 | 0 | 0 | 1 | 0 | 0 | 12 | 0 | 15 | 174 | 0 | 0 | 0 | 0 | 0 | 5 | 0 | 0 | 0 | 209 |
| Riana | 14 | 1 | 32 | 6 | 14 | 0 | 1 | 5 | 16 | 0 | 1 | 9 | 0 | 0 | 175 | 2 | 2 | 2 | 0 | 0 | 1 | 3 | 1 | 271 |
| Rigoso | 15 | 0 | 0 | 3 | 0 | 4 | 1 | 4 | 0 | 2 | 1 | 1 | 0 | 0 | 0 | 92 | 9 | 0 | 0 | 0 | 5 | 1 | 0 | 123 |
| Rimagna | 16 | 0 | 1 | 1 | 2 | 12 | 0 | 5 | 0 | 1 | 1 | 2 | 0 | 0 | 0 | 25 | 89 | 14 | 0 | 0 | 1 | 1 | 0 | 152 |
| Trefiumi | 17 | 1 | 2 | 1 | 2 | 2 | 0 | 16 | 0 | 0 | 1 | 4 | 4 | 1 | 2 | 4 | 21 | 171 | 0 | 0 | 0 | 10 | 0 | 238 |
| Trevignano | 18 | 0 | 0 | 0 | 0 | 5 | 0 | 3 | 0 | 1 | 7 | 0 | 0 | 0 | 0 | 0 | 0 | 0 | 61 | 2 | 3 | 0 | 1 | 87 |
| Vaestano | 19 | 0 | 1 | 1 | 0 | 1 | 0 | 4 | 0 | 2 | 25 | 0 | 6 | 14 | 0 | 0 | 0 | 0 | 1 | 134 | 1 | 0 | 0 | 190 |
| Valcieca | 20 | 0 | 0 | 0 | 0 | 4 | 0 | 0 | 0 | 13 | 2 | 0 | 1 | 0 | 0 | 5 | 0 | 0 | 0 | 2 | 72 | 0 | 0 | 99 |
| Valditacca | 21 | 1 | 2 | 1 | 0 | 0 | 0 | 11 | 0 | 0 | 0 | 29 | 2 | 1 | 2 | 4 | 6 | 12 | 1 | 1 | 0 | 175 | 0 | 248 |
| Villula | 22 | 0 | 1 | 23 | 2 | 0 | 1 | 0 | 1 | 0 | 0 | 0 | 0 | 0 | 0 | 0 | 0 | 0 | 0 | 0 | 0 | 0 | 76 | 104 |

TABLE 6.6. Probability for a woman of a given village to marry a man of a given village

| Wife's birthplace | Husband's birthplace | | | | | | | | | | | | | | | | | | | | | |
|---|---|---|---|---|---|---|---|---|---|---|---|---|---|---|---|---|---|---|---|---|---|---|
| | 1 | 2 | 3 | 4 | 5 | 6 | 7 | 8 | 9 | 10 | 11 | 12 | 13 | 14 | 15 | 16 | 17 | 18 | 19 | 20 | 21 | 22 |
| 1 | 917 | 917 | 935 | 938 | 938 | 977 | 979 | 995 | 995 | 996 | 996 | 996 | 996 | 998 | 999 | 999 | 999 | 999 | 999 | 999 | 999 | 999 |
| 2 | 0 | 790 | 811 | 839 | 839 | 839 | 902 | 909 | 909 | 909 | 937 | 937 | 937 | 993 | 993 | 993 | 993 | 993 | 999 | 999 | 999 | 999 |
| 3 | 20 | 22 | 892 | 902 | 903 | 916 | 919 | 934 | 934 | 940 | 940 | 940 | 940 | 945 | 945 | 945 | 945 | 945 | 945 | 945 | 948 | 999 |
| 4 | 17 | 68 | 85 | 780 | 797 | 797 | 797 | 847 | 847 | 847 | 864 | 864 | 864 | 983 | 983 | 983 | 983 | 983 | 983 | 983 | 999 | 999 |
| 5 | 0 | 11 | 11 | 11 | 770 | 770 | 853 | 853 | 868 | 917 | 921 | 921 | 932 | 932 | 940 | 955 | 962 | 962 | 966 | 992 | 999 | 999 |
| 6 | 200 | 206 | 262 | 269 | 275 | 944 | 944 | 975 | 975 | 975 | 975 | 975 | 975 | 975 | 975 | 975 | 975 | 975 | 981 | 981 | 981 | 999 |
| 7 | 0 | 35 | 58 | 61 | 117 | 117 | 878 | 880 | 880 | 892 | 930 | 930 | 936 | 945 | 950 | 965 | 971 | 971 | 974 | 977 | 999 | 999 |
| 8 | 96 | 106 | 179 | 206 | 206 | 239 | 239 | 982 | 982 | 982 | 982 | 982 | 982 | 991 | 991 | 991 | 995 | 995 | 995 | 995 | 995 | 999 |
| 9 | 0 | 0 | 0 | 0 | 54 | 54 | 54 | 54 | 914 | 925 | 925 | 935 | 935 | 935 | 946 | 946 | 946 | 946 | 968 | 999 | 999 | 999 |
| 10 | 2 | 2 | 7 | 7 | 19 | 19 | 24 | 24 | 33 | 882 | 882 | 899 | 927 | 932 | 934 | 934 | 934 | 941 | 991 | 993 | 995 | 999 |
| 11 | 4 | 18 | 29 | 29 | 33 | 33 | 143 | 143 | 143 | 143 | 766 | 766 | 766 | 780 | 806 | 828 | 850 | 853 | 853 | 853 | 999 | 999 |
| 12 | 0 | 0 | 0 | 0 | 20 | 20 | 20 | 20 | 20 | 79 | 79 | 723 | 911 | 911 | 911 | 911 | 911 | 921 | 990 | 990 | 999 | 999 |
| 13 | 5 | 5 | 10 | 10 | 10 | 10 | 14 | 14 | 14 | 72 | 72 | 144 | 976 | 976 | 976 | 976 | 976 | 976 | 999 | 999 | 999 | 999 |
| 14 | 4 | 122 | 144 | 196 | 196 | 199 | 218 | 277 | 277 | 280 | 314 | 314 | 314 | 959 | 967 | 974 | 982 | 982 | 982 | 985 | 996 | 999 |
| 15 | 0 | 0 | 24 | 24 | 57 | 65 | 98 | 98 | 114 | 122 | 130 | 130 | 130 | 130 | 878 | 951 | 951 | 951 | 951 | 992 | 999 | 999 |
| 16 | 0 | 7 | 7 | 7 | 86 | 86 | 118 | 118 | 125 | 132 | 145 | 145 | 145 | 145 | 309 | 895 | 987 | 987 | 987 | 993 | 999 | 999 |
| 17 | 4 | 13 | 17 | 25 | 34 | 34 | 101 | 101 | 101 | 105 | 122 | 122 | 126 | 134 | 151 | 239 | 958 | 958 | 958 | 958 | 999 | 999 |
| 18 | 0 | 0 | 0 | 0 | 57 | 57 | 92 | 92 | 103 | 184 | 184 | 230 | 230 | 230 | 230 | 230 | 230 | 931 | 954 | 989 | 989 | 999 |
| 19 | 0 | 5 | 11 | 11 | 16 | 16 | 37 | 37 | 47 | 179 | 179 | 211 | 284 | 284 | 284 | 284 | 284 | 289 | 995 | 999 | 999 | 999 |
| 20 | 0 | 0 | 0 | 0 | 40 | 40 | 40 | 40 | 172 | 192 | 192 | 202 | 202 | 202 | 253 | 253 | 253 | 253 | 273 | 999 | 999 | 999 |
| 21 | 4 | 12 | 16 | 16 | 16 | 16 | 60 | 60 | 60 | 60 | 177 | 185 | 190 | 198 | 214 | 238 | 286 | 290 | 294 | 294 | 269 | 999 |
| 22 | 0 | 10 | 231 | 250 | 250 | 260 | 260 | 269 | 269 | 269 | 269 | 269 | 269 | 269 | 269 | 269 | 269 | 269 | 269 | 269 | 269 | 999 |

## 6.3 THE MIGRATION MATRIX

The original migration matrix given in table 6.5 was calculated from the birthplaces of spouses given in parish marriage books for the 22 villages (here used synonymously with parishes), averaging over all data available. The average time for which information was available varied somewhat for different villages (mean number of years = 230). The original matrix shows considerable lack of symmetry, much of which is due to random variation of rare events, since the total number of marriages is not large and most elements of the matrix have zero entries. Some asymmetry, however, is part of the overall migration pattern, since there is a tendency to patrilocal marriage. This is partly masked by the custom of celebrating marriages in the bride's parish. It is possible that the migration pattern changed over the three centuries during which the demography was studied, but the numbers of marriages per village are too small for an evaluation of secular changes in the migration matrix.

Another limitation was the lack of censuses data in the upper Parma valley, while such data (*status animarum* = state of souls) were taken with some regularity in other areas. We were forced to assume that the migration pattern did not change over the whole period, an assumption that cannot be seriously incorrect for the following reasons. We have evidence that the village sizes did not change in an important way over the last three centuries, although one parish was founded in the eighteenth century and another disappeared more recently. On one hand, the total number of marriages per parish did not seem to change systematically over periods of half a century, except for a small overall decrease during the eighteenth century that was compensated for in the next one. Major changes occurred in the second half of the twentieth century, but all data collections took place before 1960. On another hand, the total number of marriages per village per year over the whole period, divided by 0.008 (a constant calculated from the number of marriages per inhabitant per year averaged over the whole area and time period) gave a village size that compared well with the village size from the 1951 census (figure 6.1, $r = 0.859$, $P << 0.001$). This distribution of village sizes was used for the initial populations in the simulation, and final values at

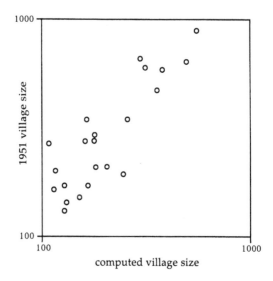

FIGURE 6.1. Correlation between computed village size averaged over time (number of marriages/0.008) and village size from the 1951 census.

the end of the simulation showed a remarkable similarity with the initial distribution. Not surprisingly, in some simulations a few villages became very small at equilibrium, and in one experiment a few disappeared. In fact, a few small villages were also much reduced in size in reality.

There are many other ways of estimating migration, and data from this area, though not or not necessarily the same, were previously analyzed by several different methods in Cavalli-Sforza (1958, 1962), Cavalli-Sforza and Bodmer (1971, 1999), Wijsman and Cavalli-Sforza (1984), and Cavalli-Sforza and Feldman (1990). We will give here a short synopsis of these results.

In another earlier sample of 21 parishes from the whole Parma valley (Cavalli-Sforza 1958) we examined factors correlated with endogamy (the proportion of marriages in which both spouses are from the same village). As shown in table 6.7, the whole Parma valley sample showed a strong positive correlation of endogamy with altitude ($r = 0.66$, $P < 0.01$), and an equally strong, inverse one with population density. By contrast, its zero-order correlation with village size was weakly positive, not significantly different from zero, but

TABLE 6.7. Endogamy (percentage of marriages in which both partners reside in the same parish) in 21 parishes from the whole Parma valley and its correlation with demographic and ecological data

| Variables | Mean | Range |
|---|---|---|
| $x$  Endogamy | 54.5% | 34–77% |
| $y$  Village size | 700 | 159–2495 |
| $w$  Population density | 86.5 | 37–80 |
| $z$  Altitude (m) | 478 | 60–950 |

| | Correlation coefficients | |
|---|---|---|
| Zero order | First order | Second order |
| $r_{xy} = +0.25$ | $\begin{cases} r_{xy.w} = +0.52* \\ r_{xy.z} = +0.67** \end{cases}$ | $r_{xy.wz} = +0.66**$ |
| $r_{xw} = -0.67**$ | $\begin{cases} r_{xw.y} = -0.76** \\ r_{xw.z} = -0.28 \end{cases}$ | $r_{xw.yz} = -0.34$ |
| $r_{xz} = +0.66**$ | $\begin{cases} r_{xz.y} = +0.82** \\ r_{xz.w} = +0.25 \end{cases}$ | $r_{xz.wy} = +0.57*$ |
| $r_{yw} = +0.20$ | | |
| $r_{yz} = -0.34$ | | |
| $r_{wz} = -0.84**$ | | |

Source. From Cavalli-Sforza (1958).
*Significantly different from zero at $P = 0.05$.
*Significantly different from zero at $P = 0.01$.

became significantly positive after partialing out the other two variables. This sample, representing the whole valley, had a considerable range of variation of all variables. By contrast, the 22 parishes of the upper Parma valley showed a much smaller range of altitude and population density (table 6.8). There is still a positive, significant zero-order correlation ($r = 0.57$, $P < 0.01$) of village size and endogamy, calculated from table 6.5. Not surprisingly, there is no change in this correlation after partialing out for altitude (or population density), because these, and especially the latter, show very little variation in the upper part of the valley. This result confirms the strong correlation of endogamy and village size.

Endogamy is only one possible (inverse) measure of migration. Another can be obtained from the whole distribution of geographic distances between birthplaces of spouses (the distribution of matri-

TABLE 6.8. Demographic and ecological data in the 22 parishes of the
upper Parma valley

| Variables | Mean | Range |
|---|---|---|
| $x$ Endogamy | 74.0% | 58–92% |
| $y$ Village size | 336 | 131–887 |
| $w$ Population density | 45.3 | 37–49 |
| $z$ Altitude (m) | 843 | 609–1131 |

Correlation coefficients

| Zero order | First order |
|---|---|
| $r_{xy} = +0.57**$ | $r_{xy.z} = +0.58**$ |
| $r_{xz} = -0.39$ | |
| $r_{yz} = -0.34$ | |

**Significantly different from zero at $P = 0.01$.

monial distances, or of matrimonial migration). In earlier evaluations
(Cavalli-Sforza 1958, 1962) the whole Parma valley data had been
subdivided into two halves: marriages from parishes above and below 400 meters of altitude. The shape of the distributions was examined with many different models (Cavalli-Sforza 1962), which could
be grouped into two major classes: diffusional and gravitational
models. The diffusion model can be conceived as the consequence of
more or less random displacement (change of location) of individuals
over their lifetime, roughly similar to Brownian motion. Gravitation
is a model in which there is individual movement only for reasons of
daily work or for going to market, fairs, or ceremonies, followed by
return to a fixed abode, home. In this second model, the only important cause of change of location is marriage, and women mostly
move unless they are the only heirs of property, in the absence of
brothers. In an essentially rural community where farming is the major occupation and land is mostly owned and inherited by fixed rules,
and there are fewer opportunities for change of location due to
change of work, a gravitational model is more accurate from a sociological point of view. This model generates a distribution of matrimonial distances that is very skewed, so that it is often well fitted by
functions $\exp(-\sqrt{\text{distance}})$. A Brownian motion model does not give
an equally good fit, unless one assumes heterogeneity of the modes
of transportation (Cavalli-Sforza 1962, Cavalli-Sforza and Bodmer

1971, 1999). This last model may turn out to be correct for the last fraction of the time covered by this investigation. In times prior to the twentieth century displacement was only on foot or by mule.

When the whole Parma valley sample was split between mountains (i.e., villages above 400 m of altitude) and plains (below 400 m), there was no major difference of the distribution of matrimonial migration (distances between birthplaces of spouses) except for the frequency of the zero distance class. This class corresponds to marriage within the village, and therefore is very similar to endogamy: its frequency was 64.2% in the mountains and 51.3% in the plains. The rest of the distribution of matrimonial distances is relatively similar in the mountains sample and in the plains. The probable explanation is that most of the movements that could lead to meeting a future spouse born in another village took place, for most of the period covered by this investigation, as all other displacements, on foot, or at most by horse or mule, and did not usually involve spending nights out. The average distance in kilometers between birthplaces of spouses observed in Parma (root mean square distance = 12.2), averaged over the whole valley, is one of the smallest recorded (Wijsman and Cavalli-Sforza 1984).

In the investigations of the whole Parma valley a further analysis was made, that of the dimensionality of the geographic distribution of the villages. It was shown that villages of the plains are distributed in a two-dimensional pattern, but in the mountains the pattern is neither one nor two dimensional. The reason is that the upper Parma valley is formed by more than one parallel valley, intercommunications among which is limited. An analysis of the dimensionality of the distribution of the mountains fraction gave a dimensionality coefficient $d = 1.54$ (calculated from the increase of the number of parishes $n$ with distance $r$ from any village using equation $n \approx r^d$; expectations for $d = 1$, $d = 2$, and $d = 1.54$ are given in table IX of Cavalli-Sforza [1962]). All authors who introduced somewhat different mathematical models of isolation by distance (see bibliography of Cavalli-Sforza and Bodmer [1971, 1999]) showed a major difference between expectations of one- and two-dimensional models, but there has been no attempt at giving expectations for an intermediate dimensionality. For this reason, it is hard to choose a model of isolation by distance for representing this population.

Migration is the major force opposing drift due to small size of the villages. In the case of the Parma valley there was more information on migration available than in most other similar investigations, making it possible to build the migration matrix among all villages on the basis of data on marriages celebrated in each village over a period of two to three centuries. This allowed us to bypass the limitations imposed by the models of isolation by distance, which are much more rigid and cannot be adapted so easily to describe real situations. When knowledge of the migration matrix is available there are two paths open for fitting the data: one is to use a direct recursion of the migration matrix (Bodmer and Cavalli-Sforza 1968, Malecot 1950) and the other is the population simulation. They should, under specific circumstances, give very similar results, except that the population simulation allows us to introduce more easily a proper consideration of the statistical variation of the number of migrants at every generation, sampling effects due to the fluctuation of the numbers of migrants, and the relatedness between sampled individuals and others.

An earlier direct comparison of observed and expected drift on the Parma data failed to take into account some possible sources of error and produced an unsatisfactory conclusion (table 8.16 in Cavalli-Sforza and Bodmer [1971, 1999]). We have repeated the comparison of expected and observed drift and consanguinity in the upper Parma valley and report it in the next two sections.


## 6.4 IS DRIFT THE ONLY CAUSE OF GENETIC VARIATION IN THE PARMA VALLEY?

Our early work showed that there is variation among villages, which is greater, the greater the altitude. Village size decreases with increasing altitude and with its population density. The spacing of villages does not change much with altitude, and migration decreases only slightly with it. Therefore, the decrease of village size must be the major cause of increased genetic drift. Decrease in migration (increasing endogamy) is of lesser magnitude. The structure of population, that is, the pattern of population distribution, can also contribute, but it will be reflected in the migration matrix. The correct evaluation of population sizes and of the migration matrix are therefore the major keys to the prediction of drift.

In the previous chapter, the evaluation of drift was done in two ways: by calculating the overall variance of gene frequencies among villages, in the form of $F_{ST}$, which allows us to eliminate the effect of the average gene frequency in different genes or alleles, and by calculating genetic distance, again as $F_{ST}$ between each village and its nearest neighbors. Both methods have confirmed the strong effect of altitude, population density, and village size, but have not supplied satisfactory comparisons with expectations. To generate expectations, it is necessary to introduce an adequate consideration of population structure. Both the migration matrix approach and that by an artificial population supply it. There remain two kinds of problems: the validity of the demographic assumptions made in both approaches. It may seem that the migration matrix requires fewer parameters than the artificial population. This is true, but there remains the problem of estimating accurately the observed variance of gene frequencies, where the demographic problems return, as, for instance, in taking account of the relatedness of individuals in the samples. The artificial population has two advantages over the migration matrix: it supplies gene frequency data that are directly comparable with those observed in the real population, and therefore less manipulation of the data is necessary. In addition, it gives direct estimates of consanguinity and inbreeding, which can also be compared with observed values, and adds further information, though of a somewhat different nature.

The first question is how to take a sample of the artificial population that is directly comparable with the observed data. The solution we employed was to take as population sample all children born in a cycle, the equivalent of ten years of simulation. This is similar to what happened in the real population where all adults of one generation, but not their children, were taken. In small villages a not insignificant fraction of sampled individuals were sibs.

All comparisons of real and artificial population were made by the same statistical parameters: chi squares among villages. These were an optimal choice for evaluating gene frequency variances, because they are already divided by $p(1 - p)$, where $p$ is the average gene frequency of the whole sample. This is exactly as is done in an $F_{ST}$ and allows comparisons between frequencies of alleles whose $p$ is not too close to 0 or 100%. The estimates of $F_{ST}$ thus obtained are very similar to those obtained by the method of Weir and Cockerham (1984) on the same data (Cavalli-Sforza and Feldman 1990).

Division of chi squares by the number of degrees of freedom (which would always contain the number of villages minus one) provides a variance, but there was no need to do it, given that the number of villages was the same in the real and artificial populations. Chi square, however, is proportional to the average number of individuals sampled per village, and it was easy to correct for this cause of difference between real and artificial populations, since this quantity is known. This correction was carried out, but it is relatively trivial, because the number of children born in a cycle is fairly close to the size of the samples of individuals whose blood was taken in the real population. It may be objected that the sample of children born in a decade has a number of sibs probably smaller than the full sibships, because the procreation of the average family lasts for about two decades. The real sample was made of adults and included some adult sibs. We earlier calculated a reduction coefficient to take this into account (Guglielmino Matessi and Zei 1979) but have realized that this is no longer necessary. The sample of the artificial population contained an approximately similar average number of sibs.

Gene frequencies were estimated by gene counting at every cycle on the children born. There were two evaluations of the variances of gene frequencies among villages. One was on the full set of alleles of each of the three gene systems; these chi squares have a number of degrees of freedom equal to the number of villages minus one, times the number of alleles minus one. The second evaluation was on each allele versus all the others of the same gene system and averaged for all the alleles; this chi square has a number of degree of freedom equal to the number of villages minus one. Here there is redundancy because all allele frequencies sum to one, but the same procedure was used on the real and the artificial population, and chi squares in the artificial and real population, with the same number of degrees of freedom and adjusted to the same numbers of individuals, were compared by $F$ ratios.

The three gene systems studied, ABO, MN, and RH, have four alleles, two alleles, and seven alleles, respectively (as in table 6.2) and thus their degrees of freedom in the comparison of the 22 vil-

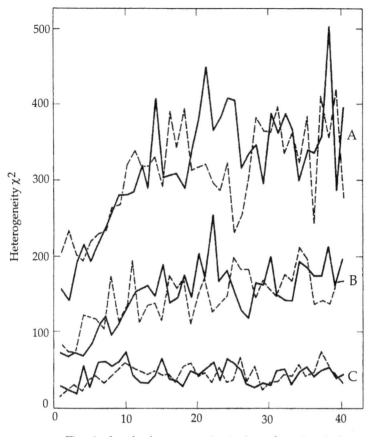

Time in decades (one generation is about three decades)

FIGURE 6.2. Heterogeneity chi square for three blood-group systems between 22 villages in an artificial population simulating the upper Parma
valley: (A) RH, (B) ABO, (C) MN. *From Cavalli-Sforza [1967].*

lages are 63, 21, and 126. Figure 6.2 shows the chi square values
after a period of 40 cycles (about 14 generations) in the first experiment. The variance of gene frequencies was at equilibrium in 6–7
generations. The chi squares of the initial generation were very close,
on average, to their degrees of freedom, as expected because at the
beginning there was no variation between villages beyond that due to
sampling from a homogeneous population. The three blood group
systems clearly increase with similar curves, displaced one from the

TABLE 6.9. Variation between gene frequencies of three gene systems in the real
and artificial population

| Gene system | Artificial population | | | Real population | | | F | |
|---|---|---|---|---|---|---|---|---|
| | df | $\chi^2$ | $\chi^2/df$[a] | df | $\chi^2$ | $\chi^2/df$ | Artificial/ real | Significance |
| ABO | 63 | 274.42 | 4.35 (3.44) | 63 | 265.09 | 4.21 | 0.82 | 23% |
| MN | 21 | 104.24 | 4.96 (3.92) | 21 | 68.06 | 3.24 | 1.21 | 33% |
| RH | 126 | 424.12 | 3.37 (2.66) | 105 | 228.96 | 2.18 | 1.22 | 13% |
| All sys- tems | 210 | 802.78 | 3.82 (3.01) | 189 | 562.11 | 2.97 | 1.02 | 44% |
| Sample size of alleles per village | | 104.20 | | | 82.36 | | | |

*Note.* Values of the variance are obtained from chi squares among villages divided by their
degrees of freedom. The last column is an *F* value calculated from the ratio of the two $\chi^2$'s
artificial/real.
[a]In parentheses are values of $\chi^2/df$ corrected for sample size.

other according to their degrees of freedom, MN being the lowest,
RH highest, and ABO intermediate.

The comparison between the real population and the artificial one
at equilibrium is shown in table 6.9. In the real population some RH
alleles were very rare, and computations were made with six alleles
only, giving 105 degrees of freedom. We still need to adjust the artificial population values to take account of the fact that the chi squares
are calculated on a higher number of alleles for the artificial population. The ratio of the average number of alleles per village, the values
of which for the real and artificial population are given at the bottom
of the two columns in table 6.10, indicate that, to make the comparison valid, chi square values of the artificial population must be multiplied by $82.36/104.20 = 0.8$. Corrected chi squares of the artificial
population are given in parentheses after the actual values, and these
corrected numbers are now expected to be equal, within statistical
error, to those of the real population. There is a general agreement
between observation and expectation, although a perfect test of goodness of fit would be very difficult, because of the redundancy of the
chi square values for the different alleles.

It would be difficult, in any case, to expect a much better agree-

TABLE 6.10. Real and artificial populations: variation between alleles of three genetic systems

| Systems | Allele | Real population $\chi^2$ per allele | Real population Average $\chi^2$ per system | Artificial population $\chi^2$ per allele | Artificial population Average $\chi^2$ per system |
|---|---|---|---|---|---|
| ABO | $A_1$ | 64.76 | | 120.96 | |
| | $A_2$ | 54.65 | 92.06 | 45.82 | 91.98 |
| | B | 127.23 | | 113.14 | |
| | O | 121.61 | | 88.00 | |
| MN | M | 68.05 | 68.05 | 104.24 | 104.24 |
| RH | r | 69.72 | | 61.92 | |
| | $R_1$ | 54.45 | | 51.52 | |
| | $R_2$ | 45.01 | 48.01 | 102.36 | 68.51 |
| | $R_0$ | 30.04 | | 48.50 | |
| | R' | 14.44 | | 72.48 | |
| | Other | 74.37 | | 74.30 | |
| Average $\chi^2$ | | 65.85 | | 80.29 | |
| SD | | 34.04 | | 27.04 | |
| Sample size of alleles per village | | 82.36 | | 104.20 | |
| Ratio between sample sizes | | | 0.8 | | |

ment, but we want to comment on the less good agreement shown by the ABO genes, the only system that shows higher chi squares between villages in the real population than in the artificial population. This discrepancy is more clearly seen when the mean chi squares per allele for each genetic system are computed (table 6.10). We could trace a very special reason why ABO behaves as an outlier.

An examination of the numbers of individuals of the four alleles in different villages shows that one particular village, Rimagna, is an outlier for the B gene (see table 5.7), having an inordinately large number of B subjects. Gene O is especially low. This is not true of any of the neighboring villages. No other equally extreme outlier was observed. It is possible that an extreme demographic bottleneck occurred at some time in Rimagna. The name suggests a Germanic origin (*Arimannia* is a Longobardic word) but more than a millennium has passed since there were settlements of Goths and Longobards in this area, and it is very unlikely that a founder effect, even if very strong, would survive this amount of time in the presence of a reasonable gene flow from neighboring villages. Moreover, Rimagna's endogamy in the last three centuries (89/152 = 0.586 in table 6.6) is very low. Individuals of different ABO genotypes show considerable differences in susceptibility to infectious diseases (Mourant et al. 1958) and it is possible that an epidemic episode for one of them caused the peak of B genes, with loss of O's. Another reason to think in terms of an episode of natural selection against the O blood group is that drift would tend to affect other genes similarly, but the other genes tested do not show abnormal frequencies in Rimagna. It is, of course, difficult to consider the available evidence sufficient to affirm that natural selection rather than drift is responsible for this modest episode of variation.

We also tested for the effect of socioeconomic conditions on drift. Table 6.11 shows the values of the mean chi squares per alleles obtained in seven artificial population experiments. In four of these, marital choice for socioeconomic conditions was introduced, with a medium correlation between spouses ($r = 0.4$) in two of them (experiments 3 and 6) and a high correlation ($r = 0.8$) in the other two (experiments 4 and 7). In the artificial population socioeconomic conditions were inherited by one sex only. As the table shows, there was no effect of variation for socioeconomic conditions on genetic drift. All chi squares are calculated at the 40th cycle, when equilibrium was reached for all. Drift is perfectly comparable for the three gene systems, which give essentially the same variation.

As also described in our earlier papers, in the lower part of the valley there was no significant variation among villages. Much larger samples of individuals than those examined here would be necessary for detecting the small amount of drift expected in the areas like the plains, where villages are larger. Here drift could cause only a minor excess in the variation of gene frequencies over that expected under completely random sampling of a totally homogeneous population.

The total absence of observable variation between villages in the lower Parma valley, where population density and village size are greater by a factor of 2 or more, allows us to extend the conclusion to

TABLE 6.11. Artificial population: variation between alleles of three genetic systems in 7 experiments

| Experiment no. | Social condition | $\chi^2$ per allele | | | | Sample size of alleles per village |
| --- | --- | --- | --- | --- | --- | --- |
| | | ABO | MN | RH | Average | |
| 1 | 0 | 91.98 | 104.24 | 69.34 | 88.52 | 104.2 |
| 2 | 0 | 126.20 | 148.76 | 97.54 | 124.17 | 105.8 |
| 3 | 0.4 | 129.34 | 89.96 | 133.72 | 117.67 | 97.2 |
| 4 | 0.8 | 102.74 | 158.66 | 101.66 | 121.02 | 104.8 |
| 5 | 0 | 107.08 | 88.72 | 113.92 | 103.24 | 100.0 |
| 6 | 0.4 | 115.80 | 70.86 | 92.86 | 93.17 | 101.2 |
| 7 | 0.8 | 135.76 | 79.64 | 98.76 | 104.72 | 99.8 |
| Average ± SD | | 115.56 ± 41.88 | 105.83 ± 90.93 | 101.11 ± 52.07 | 107.50 ± 61.63 | 101.86 ± 8.33 |

the whole Parma valley and to confirm that the difference in village size (and in population density) of the plains and mountains is the main reason why drift is observed only in the mountains, where villages are smaller, while migration remains approximately the same.

## 6.5 EXPECTED AND OBSERVED CONSANGUINITY

Around the 25th cycle the population could be considered in equilibrium as far as drift was concerned. It was assumed, without proof, that it was also in equilibrium for inbreeding, and starting with the 25th cycle until the 40th all marriages taking place in the artificial population (almost 400 per cycle) were recorded, so as to be able to estimate consanguinity observed in the population. The marriages were to be used for comparisons with the real population, as an estimate of consanguinity expected in a population in which marriage occurs on the basis of preference for age, social condition, and place of origin of mates, but independently of their degree of relationship. Incestuous unions (between sibs, parent, and children) were not allowed to take place in the artificial population.

It was easy to establish the type and degree of consanguinity of each marriage of the artificial population, because every individual born had a progressive identification number, IN, and carried the INs of his or her father and mother and of all his or her children, making it easy to establish genealogies, descending or ascending in time. A genealogy of an individual going back to his or her great-great-grandparents is shown in figure 6.3. Common ancestors are 35690, 35331 (maternal great-great-grandparents and paternal great-grandparents), 31011 (paternal great-great-grandparent but also maternal great-grandparent), who married twice (to 30632 and 33363), 38043, and 38980 (paternal and maternal grandparents). The figure marks differently common and intermediate ancestors. Children of husband 47203 and wife 47172 are first cousins, 2½ cousins, and half-2½ cousins. The inbreeding coefficient is $F = 0.0742$ (see also Cavalli-Sforza and Bodmer [1971], figures 7.2 and 7.3, Table 7.2).

A sample of 1,000 marriages was studied and results of the analysis of their consanguinity is compared in table 6.12 with the frequencies of consanguineous marriages celebrated in the 22 villages

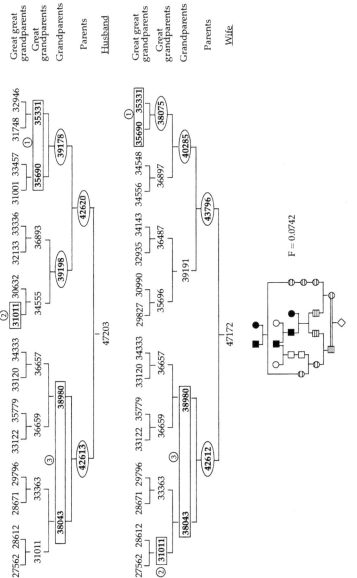

FIGURE 6.3. Genealogy of an individual going back to his great-great-grandparents in an artificial population. Common ancestors are in rectangles; intermediate ancestors are in ovals. At the bottom, a stylized representation of the tree: common ancestors are in black; intermediate ancestors are lined.

TABLE 6.12. Consanguinity in artificial population and in real population (% of total marriages)

| | Artificial population | | | | | | | Real population | | |
|---|---|---|---|---|---|---|---|---|---|---|
| | Single | | | Multiple | | | Total | Single | Multiple | Total |
| | Full | Half | Total | Full | Half | Total | | | | |
| Uncle–niece | 0.1 | — | 0.1 | 0.1 | — | 0.1 | 0.2 | 0.10 | 0.02 | 0.12 |
| First cousins | 1.6 | — | 1.6 | 2.0 | 0.2 | 2.2 | 3.8 | 3.65 | 0.30 | 3.95 |
| First, once removed | 1.7 | 0.4 | 2.1 | 1.0 | 0.7 | 1.7 | 3.8 | 1.49 | 0.24 | 1.73 |
| First, twice removed | 0.2 | 0.2 | 0.4 | — | — | — | 0.4 | — | — | — |
| Second cousins | 3.4 | 0.2 | 3.6 | 4.0 | 0.5 | 4.5 | 8.1 | 8.13 | 0.68 | 8.81 |
| Second, once removed | 3.6 | 0.5 | 4.1 | 4.6 | 1.9 | 6.5 | 10.6 | 1.40 | 0.16 | 1.56 |
| Third cousins | 7.5 | 0.8 | 8.3 | 13.3 | 1.5 | 14.8 | 23.1 | 2.18 | 0.56 | 2.74 |
| Total marriages | | | | | | | | | | |
| % single consanguinity (N) | | | 20.20 (202) | | | | | | | 16.96 (970) |
| % multiple consanguinity (N) | | | 12.50 (125) | | | | | | | 0.98 (56) |
| Average inbreeding coefficient, α ‰ | | | 6.443 | | | | | | | 4.769 |
| α, up to and including 2nd cousins | | | 4.797 | | | | | | | 4.536 |

Note. Multiple consanguinity marriages are counted as many times as the simple consanguinities they contain.

of the real population over the 1850–1950 period. The differences between the frequency of consanguinity in real and artificial populations are clear-cut. The total number of consanguineous marriages (single + multiple) is significantly higher in the artificial (32.7%) than in the real (17.9%) population. The most consistent differences are as follows:

1. The frequencies of most remote single-cousin marriages (second cousins once removed and third cousins) are about three times higher in the artificial than in the real population. This major discrepancy requires a further qualification. In the real population the low frequency of remote cousin marriages is in part due to the suspension of dispensations after the year 1917. But even before that date it was always low. The ratio between remote and recent consanguinity (including second cousins) is 1.6 in the artificial population and 0.3 in the real one (in the Parma diocese). There may be greater difficulty for parish priests to detect remote rather than recent consanguinity when it is not declared by the families of future mates.

2. Multiple consanguinity, relative to simple consanguinity, is much smaller in the real than in the artificial population. There is a strong difference in the ratio simple/multiple consanguinity in the artificial and the real population for all marriages (1.62 vs. 17.32). It is useful to examine separately the two types of consanguinity.

A flaw of the artificial population is probably responsible for a slight paradox: simple consanguineous marriages of uneven degree (uncle–niece, first cousins once removed, first cousins twice removed) are relatively more frequent in the artificial population, and those of even degree (first and second cousins) are more frequent in the real population. Considering the small difference, and the cost of running simulations, this flaw was not considered important enough to require a repetition of the experiment. In the first case, the difference is significant only for 1½ cousins; the explanation is to be sought most probably in the excessive value in the simulation of the variance of male generation time. The comparison of demographic data in the real and the artificial population (after 40 cycles) is shown in table 6.13.

TABLE 6.13. Demographic data of artificial and real population

|  | Artificial population | Real population |
|---|---|---|
| Generation time |  |  |
| Females |  |  |
| Mean | 29.5 | 30.9 |
| Variance | 43.4 | 42.6 |
| Males |  |  |
| Mean | 35.7 | 36.1 |
| Variance | 81.0 | 53.3 |
| Mean age at marriage |  |  |
| Females | 24.4 | 25.8 |
| Males | 30.7 | 30.6 |
| Variance of age difference |  |  |
| Between mates | 54.1 | 53.1 |
| Between sibs | 39.8 | 33.3 |

The second case (greater frequency of first and second cousins in the real population compared with the simulated one, both very highly significant) is most probably explained by considering simultaneously single and multiple consanguinity. Multiple-consanguinity marriages can all be identified in the simulation, and counted as many times (from 2 to 5) as the simple consanguinities they contain (table 6.12). In reality, such a rigorous identification in the real situation is almost impossible. The small number of multiple consanguinity marriages observed in the real population are only of "duplex" type: double first cousins, first and second cousins, double second cousins, and so on. It seems very likely that parish priests were usually content to discover the most recent consanguineous bonds, which are the most important ones, and did not consider it necessary to describe all existing consanguinities. If this is correct, a substantial number of multiple consanguinities were recorded as simple consanguinity. First or second cousins would therefore appear more frequent than expected. But the lost, more remote consanguinities contribute little to the inbreeding coefficient, the underestimate of which is therefore minor.

In transforming the excess of first- and second-cousin marriages in the real population into multiple consanguinities, the frequency of the latter would increase to 4.3%, a value that is still too low if compared with that of the artificial population (12.5%). This procedure would still not detect all the losses. But it is interesting that if one sums the single and multiple consanguinities of first cousins, and, separately, that of second cousins, the frequencies in the artificial and real population are remarkably similar.

The final result in terms of the average inbreeding coefficient is still higher in the artificial population (6.44 per thousand) versus the real one (4.77 per thousand), for a variety of reasons; the most important ones appear to be the complete detection of more remote consanguinities in the case of multiple ones, and the more complete survey of remote consanguinities (2½ and third cousins). When one eliminates from consideration these more remote cousins, the agreement of the average inbreeding coefficient is much higher in the two populations (4.80 per thousand in the artificial population and 4.54 per thousand in the real one). There may still be a small avoidance of consanguineous marriages, but it is certainly minor.

Results of the simulation allow us to conclude that the loss of inbreeding in the real population, due to avoidance of consanguineous marriages, is lower than was anticipated on the basis of the analysis conducted in the earlier chapters. The number of dispensations requested (if not that of marriages actually occurring, which is not known) is not very far from that which one would expect if consanguinity was not a factor in the decision on choice of a spouse, but other major marriage determinants like age and location of the potential spouses are given adequate consideration. The superficial differences between the real and artificial populations seem to be amply resolved, considering that, most probably, in the real population the most remote consanguinities in the very frequent cases of multiple consanguinity were not detected or not indicated. This seems a reasonable error, which we conclude was made frequently by parish priests, but, especially if recognized, does not detract from the value of consanguinity dispensations for reconstructing inbreeding in populations. Of course, this does not consider the losses due to the limitations of consanguinity detected in dispensations, caused by the exclusion of more remote consanguinities.

The conclusion is that in the populations we have studied there is no real or major avoidance of consanguineous marriages, and that they almost happen at random as if they were entirely allowed. It is worth, however, to remember that the data from bishoprics are based on requests of dispensation, and not on marriages that actually occurred.

Although there is a fee to be paid for requesting a dispensation, the poor do not have to pay—and poverty is sometimes a cause for granting dispensations. Parish priests seemed to rarely refuse to transmit the dispensation to the bishop, if the consanguinity degree is of acceptable level. The fraction of dispensations that were refused is not known but we have evidence indicating that it is not high, as the Vatican data show a reasonable correspondence with the data of dioceses of origin for corresponding periods. But the time passing between the request of dispensations by the prospective spouses and the arrival of the dispensation was probably sufficiently long in older times and there might have been marriages that did not take place because one or the other of the spouses changed his or her mind or died in the interval. This fraction cannot be large.

CHAPTER 7

# Islands

## 7.1 ITALIAN ISLANDS

The Italian Republic has a territory of a little over 301,000 km$^2$. This includes 25,400 km$^2$ of the island of Sicily (the largest Mediterranean island), plus the Aeolian, Egad, and Pelagian archipelagos, and Ustica and the Pantelleria islands, totaling about 280 km$^2$; as well as 23,800 km$^2$ of the island of Sardinia, the next largest, near which are five small islands totaling 256 km$^2$. Because of their insular status Sicily and Sardinia have been under a special regime concerning consanguinity dispensations, and therefore they are analyzed separately from the rest of Italy. Sardinia is the better known and is considered first. Sicily could be examined only in part and is considered next. Another important difference between these two major Mediterranean islands is that Sicily is very close to the Italian peninsula and in the remote past was also much closer to Tunisia, especially during the last glacial age, when the sea level was much lower than now. Sardinia, on the other hand, is rather isolated from the mainland. The most important group of small islands, Aeolian or Lipari, are considered next.

## 7.2 SARDINIA

Sardinia is close to another island, Corsica, the third largest in the Mediterranean. Corsica is linguistically and geographically, but not politically, Italian. Genetically the two islands are rather distinct, with Corsica closer to the mainland, while Sardinia has unique genetic characteristics (Francalacci et al. 2003). Though still clearly European, Sardinians form the most distinct isolate in the European continent after the Saami (also called Lapps, a derogatory name) who live in northern Scandinavia and speak a non-Indo-European language); they are more distinct than, say, the Basque, who also speak a non-Indoeuropean language (Cavalli-Sforza et al. 1994). The language spoken in Sardinia is an Italian dialect that still shows strong Latin influences.

The genetic distinctness of Sardinians speaks for a marked isolation following the first known settlement, which happened in Paleolithic times at least 11,000 years ago (Spoor and Sondaar 1986). In the Bronze Age, which began here around 3800 years ago, there was a period of construction of stone towers (called *nuraghi*) that were probably at the center of small villages. Similar "megalithic" cultures are common in the Mediterranean and also outside it on the Atlantic coast of Europe, but the shape of the *nuraghi* is characteristic and almost unique to Sardinia. Placed in a net at a regular distance between each other as well as from the coast, they were probably used for communication among villages and as fortresses. Their population at the peak may have been of the order of 300,000 (Lilliu 1983) which is almost one-fifth of those living in Sardinia now, and more than those living there at the end of the seventeenth century (see table 7.1).

In the first millennium B.C. new settlers from Tunisia (Phenicians and later Carthaginians, themselves from a Phenician colony in Tunisia) built a number of colonies on the Sardinian coast, especially in the southern part of the island, and established contact with the *nuraghi* people. The island is largely mountainous, with the major mountains in the center-east, and lends itself to a mostly pastoral economy, with agriculture mostly near the coast.

Not far from the sea, however, malaria has taken a severe toll, starting probably during the Roman occupation, which began in 238 B.C. Malaria could be controlled in Sardinia only after World War II (Bruce-Chwatt and Zulueta 1980) thanks to a campaign of malaria eradication started by the Rockefeller Foundation with ample spraying of DDT. The campaign was unsuccessful, in the sense that the initial hope of eradicating the population of mosquito vectors proved unfeasible. It was only possible to cause a retreat of mosquitoes to areas that were not densely populated by humans, but, since the survival of the malaria parasite requires adequate exchange between hu-

TABLE 7.1. Population of Sardinia from 1688 to 1991 per altitude region

| Census | Mountains | Altitude regions | | Plains | Sardinia |
|---|---|---|---|---|---|
| | | Internal hills | Coastal hills | | |
| 1688 | 18,279 | 104,327 | 31,469 | 75,457 | 229,532 |
| 1698 | 20,891 | 115,408 | 36,852 | 86,006 | 259,157 |
| 1728 | 23,844 | 137,812 | 45,173 | 105,073 | 311,902 |
| 1751 | 34,081 | 160,620 | 57,930 | 108,174 | 360,805 |
| 1771 | 32,996 | 159,079 | 59,830 | 108,880 | 360,785 |
| 1776 | 39,702 | 182,818 | 71,964 | 128,163 | 422,647 |
| 1781 | 38,396 | 190,033 | 73,658 | 129,810 | 431,897 |
| 1821 | 38,458 | 194,402 | 82,583 | 146,488 | 461,931 |
| 1824 | 37,729 | 198,707 | 85,038 | 148,357 | 469,831 |
| 1838 | 40,891 | 219,084 | 97,419 | 168,091 | 525,485 |
| 1844 | 43,421 | 224,336 | 101,667 | 174,829 | 544,253 |
| 1848 | 43,458 | 229,126 | 104,852 | 177,281 | 554,717 |
| 1857 | 42,323 | 236,107 | 109,632 | 185,181 | 573,243 |
| 1861 | 45,936 | 247,801 | 117,119 | 198,159 | 609,015 |
| 1871 | 49,057 | 256,965 | 124,387 | 206,004 | 636,413 |
| 1881 | 50,975 | 269,163 | 139,200 | 221,112 | 680,450 |
| 1901 | 61,656 | 296,221 | 187,226 | 250,690 | 795,793 |
| 1911 | 68,332 | 314,858 | 204,780 | 280,211 | 868,181 |
| 1921 | 70,643 | 316,388 | 205,955 | 292,481 | 885,467 |
| 1931 | 75,714 | 337,964 | 229,148 | 340,934 | 983,760 |
| 1936 | 78,185 | 351,326 | 236,076 | 368,619 | 1,034,206 |
| 1951 | 86,103 | 401,844 | 316,076 | 472,000 | 1,276,023 |
| 1961 | 92,404 | 416,813 | 333,493 | 576,652 | 1,419,362 |
| 1971 | 80,708 | 388,877 | 334,970 | 669,245 | 1,473,800 |
| 1981 | 76,709 | 395,754 | 370,384 | 751,328 | 1,594,175 |
| 1991 | 70,718 | 394,035 | 398,168 | 785,327 | 1,648,248 |

mans and mosquitoes, the limitation of contact between them was enough to cause the disappearance of the disease. The population of the lowlands is affected by some of the highest frequencies of β-thalassemia, a recessive anemia that, like sickle cell anemia, gives resistance to *Plasmodium falciparum* malaria in the heterozygous state, and also of G6PD (glucose-6-phosphatase deficiency), which also confers some protection against the same parasite (Siniscalco et al. 1961).

At the end of the western Roman Empire Sardinia suffered a temporary invasion of Vandals and was then occupied for some centuries by Byzantines (the eastern Roman Empire). In the last millennium there were some new settlers, especially in the northwest (from Catalonia) and in the northeast (from Pisa, Tuscany, and, to a lesser extent, Genoa). These are reflected in the local dialects and languages, but the major foreign influence and population increase was felt in the south, especially in the city of Cagliari, the major port, and in the whole southern coast. Since 1720 Sardinia became part of the kingdom of Savoy, with its capital in Turin, north Italy.

The population censused since 1688 and divided by altitude regions is given in table 7.1 (from Angioni et al. 1997). At the beginning of censuses, population was particularly low but started growing. Between 1688 and 1900 increased 3 to 6 times, according to altitude region. Since 1900 it has grown almost only in the plains and coastal regions, especially rapidly after the end of malaria (around 1950).

Consanguinity frequencies in this period, collected by Moroni and collaborators (1972), are given in table 7.2 (absolute and relative frequencies). An increase of consanguinity begins in the nineteenth century, slowly at first and then faster, especially for first cousins. It peaks in the period between the two world wars and then starts a fast descent. There is a strong effect of altitude on the frequency of consanguineous marriages, as usual, and some difference in the dynamics of the phenomenon also according to the main types of cousins. Figures 7.1 and 7.2 show the change with time in the 1850–1970 period of relative frequencies of the two major types of consanguineous matings in the four major environments. As usual, the plains have the lowest frequencies of consanguineous mates, and the coastal hills are next, with median frequency about twice as high as the plains. Internal hills are about three times as high, and mountains (located in the eastern central region) have a frequency five times higher. The data follow reasonably well the usual inverse relationship of consanguinity frequency with population density (table 7.3).

On average, first cousins and second cousins show the same behavior, but the dynamics in time is quite different. First cousins tend to peak somewhat before 1930, perhaps slightly later than in northern Emilia, while second cousins are almost constant in time, except for the mountains, which have a peak at the same time as first cousins. Results are quite similar to those observed in northern Emilia, though average inbreeding coefficients tend to be slightly higher in Sardinia.

The analysis of the frequencies of pedigree types only partially repeats the conclusions reached in northern Emilia. With regard to age difference between husband and wife, in first cousins (table 7.5) pedigree types with $\Delta = d$ are favored, as expected, while in second (table 7.7) and third cousins (table 7.9) pedigrees with $\Delta = 0$ show the maximum frequency, as also happens in northern Emilia. The asymmetric pedigrees (tables 7.4, 7.6, and 7.8) show a strong effect of age difference in the expected direction.

With regard to the number of males among the intermediate ancestors, in the first cousins the maximum frequency of pedigrees is in the class of zero males. The simplest interpretation of this phenomenon is, as mentioned in chapter 3, the importance of mothers of consanguineous mates in arranging marriages in the family and/or keeping family ties. Figures 7.3 and 7.4 show a somewhat anomalous frequency pattern for asymmetric pedigrees, which seems to be inverted compared with that observed in northern Emilia. Moreover, for the number of male ancestors, an ambiguous pattern, being intermediate between the expected one on the basis of northern Emilia observations and that observed in Sardinia for asymmetric pedigrees, is shown by second and third cousins.

Figure 7.5 summarizes the effect of the number of males as intermediate ancestors on the frequency of consanguineous marriages in even cousins, which are less affected than odd cousins by the overpowering effects of age difference. If the only factor were a stronger tendency of females than males to relocate to join husbands at marriage, these curves should show a regular increase of the frequency of marriages with an increasing number of males. In northern Emilia there were only less serious deviations from this hypothesis, but clearly in Sardinia this explanation is not sufficient.

One possible explanation is that in part of the area the change of residence of the spouses is the opposite to that observed in Emilia. We therefore see an average of two curves, one of them showing a nonlinear increase of marriages with the number of males (because males migrate less than females at marriage and therefore patrilinear descendants are more likely to remain in the same area than matrilinear ones), while the other shows a nonlinear decrease (because females migrate less than males at marriage). The greater ease of detecting consanguineous marriages with a great number of males may also contribute to increase the relative importance of the first

TABLE 7.2. Frequencies of consanguineous marriages in Sardinia

| Years | Number of consanguineous marriages | | | | | | | Total marriages |
|---|---|---|---|---|---|---|---|---|
| | 12 | 22 | 23 | 33 | 34 | 44 | Multiple | |
| 1765–1849 | 0 | 171 | 198 | 1,101 | 1,324 | 2,861 | 134 | 96,042 |
| 1850–59 | 4 | 170 | 158 | 559 | 534 | 1,008 | 267 | 28,188 |
| 1860–69 | 6 | 271 | 248 | 771 | 641 | 1,102 | 276 | 30,538 |
| 1870–79 | 16 | 429 | 305 | 1,053 | 772 | 1,314 | 333 | 36,999 |
| 1880–89 | 16 | 509 | 351 | 1,146 | 710 | 1,213 | 333 | 40,665 |
| 1890–99 | 10 | 543 | 277 | 1,038 | 590 | 983 | 286 | 42,164 |
| 1900–09 | 15 | 698 | 293 | 1,041 | 560 | 931 | 317 | 50,047 |
| 1910–19 | 10 | 1,053 | 371 | 980 | 318 | 566 | 262 | 52,729 |
| 1920–29 | 30 | 1,725 | 458 | 1,500 | — | — | 127 | 69,525 |
| 1930–39 | 19 | 1,480 | 457 | 1,580 | — | — | 97 | 72,354 |
| 1940–49 | 16 | 1,285 | 451 | 1,688 | — | — | 115 | 76,270 |
| 1950–59 | 29 | 1,070 | 422 | 1,558 | — | — | 99 | 84,100 |
| 1960–69 | 8 | 512 | 190 | 867 | — | — | 45 | 97,588 |

Consanguineous marriages ‰

| Years | 12 | 22 | 23 | 33 | 34 | 44 | Multiple | Total marriages % | α‰ |
|---|---|---|---|---|---|---|---|---|---|
| 1765–1849 | 0 | 1.78 | 2.06 | 11.46 | 13.79 | 29.80 | 1.40 | 6.03 | 0.61 |
| 1850–59 | 0.14 | 6.03 | 5.60 | 19.83 | 18.94 | 35.76 | 9.47 | 9.58 | 1.64 |
| 1860–69 | 0.20 | 8.87 | 8.12 | 25.25 | 20.99 | 36.09 | 9.04 | 10.86 | 1.82 |
| 1870–79 | 0.43 | 11.59 | 8.24 | 28.46 | 20.87 | 35.51 | 9.00 | 11.41 | 2.06 |
| 1880–89 | 0.39 | 12.52 | 8.63 | 28.18 | 17.46 | 29.83 | 8.19 | 10.52 | 2.08 |
| 1890–99 | 0.24 | 12.88 | 6.57 | 24.62 | 13.99 | 23.31 | 6.78 | 8.84 | 1.86 |
| 1900–09 | 0.30 | 13.95 | 5.85 | 20.80 | 11.19 | 18.60 | 6.33 | 7.70 | 1.61 |
| 1910–19 | 0.19 | 19.97 | 7.04 | 18.59 | 6.03 | 10.73 | 4.97 | 6.75 | 2.09 |
| 1920–29 | 0.43 | 24.81 | 6.59 | 21.57 | — | — | 1.83 | 5.52 | 2.28 |
| 1930–39 | 0.26 | 20.45 | 6.32 | 21.84 | — | — | 1.34 | 5.02 | 1.94 |
| 1940–49 | 0.21 | 16.85 | 5.91 | 22.13 | — | — | 1.51 | 4.66 | 1.71 |
| 1950–59 | 0.34 | 12.72 | 5.02 | 18.53 | — | — | 1.18 | 3.78 | 1.35 |
| 1960–69 | 0.08 | 5.25 | 1.95 | 8.88 | — | — | 0.46 | 1.66 | 0.57 |

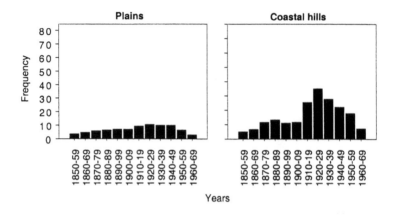

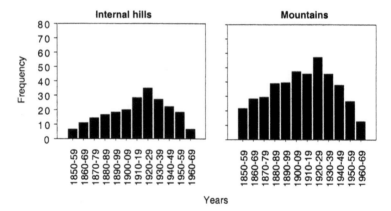

FIGURE 7.1. Sardinia 1850–1970. Average frequency of first-cousin marriages in four ecological areas.

curve, but the only class expected to give a strong contribution is that in which all intermediate ancestors are males, because it is the only one in which the two consanguineous mates have the same surname (see also chapter 4, section 3 for a possible explanation of the high frequency of this type of pedigrees).

It is possible that the differences in behavior are correlated with occupation or with customs, and that one or the other or both vary with the geographic area. It is therefore useful to subdivide the area and look for heterogeneity in the curves if numbers remain sufficient to obtain significant conclusions after subdivision of the data.

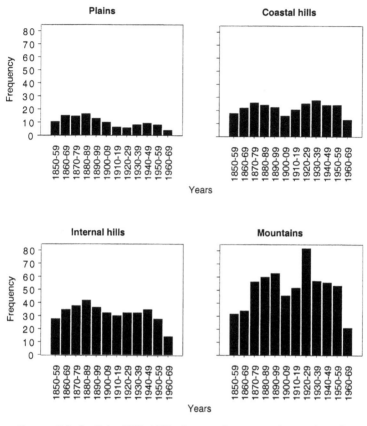

FIGURE 7.2. Sardinia 1850–1970. Average frequency of second-cousin marriages in four ecological areas.

The four areas most apt for representing the ecological and likely ethological variation are the altitude zones already considered: plains, coastal hills, internal hills, and mountains. The fourth zone can be usefully split into two: a northern area including the southern part of the Sassari province and the northern part of the Nuoro province, and a southern area limited to the southern part of the Nuoro province, which includes the center-east of the island and corresponds to the region called Barbagia (Civitates Barbariae or cities of the barbarian population, according the Roman designation). Barbagia is the most highly conservative and archaic part of the island, as shown also by the study of language and toponyms.

TABLE 7.3. Sardinia: population density in the four major environments

|  | Population density | Consanguineous marriages % 1850–1970 |
|---|---|---|
| Plains | 107.1 | 2.58 |
| Coastal hills | 43.9 | 6.33 |
| Internal hills | 43.5 | 9.58 |
| Mountains | 26.0 | 15.82 |

Second cousins are the most stable in time (figure 7.2) and therefore are more reliable indicators of other trends. In figure 7.6 they show significant variation of the curves connecting average frequency of the marriages as a function of the number of males among intermediate ancestors, distinguished according to five ecological/ethological areas. In table 7.10 the ratios between the extreme numbers of males as intermediate ancestors, 0 and 4, are compared. Their ratio varies systematically with altitude, from a value of circa 1.4 to less than 0.9.

It seems likely that the two major types of economy, agricultural and pastoral, the ratio of which varies considerably in the various areas, is responsible of this difference in behavior of migration at marriage. In the plains and coastal hills agriculture is prevalent, and males tend to stay in the area where they own the land (patrilocal

TABLE 7.4. Sardinia (1800–1970)—uncle–niece and aunt–nephew: matrix of types of pedigrees

| $\Delta$ | $K$ | | | Total | Mean frequency |
|---|---|---|---|---|---|
|  | 0 | 1 | 2 |  |  |
| Aunt–nephew |  |  |  |  |  |
| $-h$ | — | 2 : 7 | — | 7 | **7.00** |
| $-h + d\,(-w)$ | 0 : 10 | — | — | 10 | **10.00** |
| Uncle–niece |  |  |  |  |  |
| $h - d\,(w)$ | — | 1 : 65 | — | 65 | **65.00** |
| $h$ | — | — | 3 : 40 | 40 | **40.00** |
| Total | 10 | 72 | 40 | 122 |  |
| Mean frequency | **10.00** | **36.00** | **40.00** |  | **30.50** |

*Note.* Rows are categories of expected age difference, $\Delta$; columns are male numbers among intermediate ancestors, $k$. Code of pedigree in italic type, frequency in roman type. Numbers in bold are the average frequencies of each cell of the matrix.

TABLE 7.5. Sardinia (1800–1970)—first cousins:
matrix of types of pedigrees

| Δ | K 0 | 1 | 2 | Total | Mean frequency |
|---|---|---|---|---|---|
| −d | — | 1 : **1,334** | — | 1,334 | **1,334.00** |
| 0 | 0 : **2,155** | — | 3 : **1,890** | 4,045 | **2,022.50** |
| d | — | 2 : **2,153** | — | 2,153 | **2,153.00** |
| Total | 2,155 | 3,487 | 1,890 | 7,532 | |
| Mean frequency | **2,155.00** | **1,743.50** | **1,890.00** | | **1,883.00** |

*Note.* Rows are categories of expected age difference, Δ; columns are male numbers among intermediate ancestors, *k*. Code of pedigree in italic type, frequency in roman type. Numbers in bold are the average frequencies of each cell of the matrix.

migration). But in the mountains, especially in Barbagia, where pastoral activities prevail, males leave the village to take their herds to the plains in the winter (transhumance), but tend to marry in the village of origin, where their wives remain during the winter (matrilocal migration). (For an exhaustive exposition of the customs, and their changes in time, of the pastoral family in Sardinia, see Murru Corriga [1990]).

The contrast farmers/pastors became highly significant (table 7.11), taking the two extreme ecological classes. The two types of economy do not seem to influence the behavior of first cousins, which remains essentially the same in the various areas. Third cousins are too few to show significant variation.

It is also possible, considering the peak of consanguineous marriages in time followed by a descent, that the heterogeneity of customs has taken place over time rather than in space. One explanation for the increase is the desire to avoid fragmentation of land due to changes of inheritance rules, in particular, the loss of the custom of primogeniture with the Napoleonic code at the beginning of the nineteenth century. This explanation has been suggested by Moroni (see Moroni et al. 1992).

Sardinia was not affected by the Napoleonic code, but the law of the "chiudende" (1820), which consented to close the "terre comunali" (analogs of the "commons") and make them available for farming, may have had some effect, at least in the wealthier classes.

TABLE 7.6. Sardinia (1800–1970)—1½ cousins: matrix of types of pedigrees

| Δ | K 0 | 1 | 2 | 3 | Total | Mean frequency |
|---|---|---|---|---|---|---|
| **Bride in the shorter branch** | | | | | | |
| −h − d | — | — | 10 : 21 | — | 21 | **21.00** |
| | | 2 : 51 | | | | |
| −h | — | 8 : 34 **42.5** | — | 11 : 45 **45** | 130 | **43.33** |
| | | | 3 : 118 | | | |
| −h + d | 0 : 82 **82** | — | 9 : 91 **104.5** | — | 291 | **97.00** |
| −h + 2d | — | 1 : 195 | — | | 195 | **195.00** |
| Total | 82 | 280 | 230 | 45 | 637 | |
| Mean frequency | **82.00** | **93.33** | **76.67** | **45.00** | | **79.62** |
| **Groom in the shorter branch** | | | | | | |
| h − 2d | — | 5 : 493 | — | — | 493 | **493.00** |
| | | | 7 : 343 | | | |
| h − d | 4 : 356 **356** | — | 13 : 305 **324.0** | — | 1004 | **334.67** |
| | | 6 : 164 | | | | |
| h | — | 12 : 177 **170.5** | — | 15 : 180 **180** | 521 | **173.67** |
| h + d | — | — | 14 : 80 | — | 80 | **80.00** |
| Total | 356 | 834 | 728 | 180 | 2098 | |
| Mean frequency | **356.00** | **278.00** | **242.67** | **180.00** | | **262.25** |

*Note.* Rows are categories of expected age difference, Δ; columns are male numbers among intermediate ancestors, $k$. Code of pedigree in italic type, frequency in roman type. Numbers in bold are the average frequencies of each cell of the matrix.

The abolition of the feudal system (1836–1839) made it possible to share or sell small properties, especially in mountainous areas, while in the more fertile plains it caused no major changes in the distribution of land property, explaining the lesser increase in first cousin marriages.

After Napoleon's fall there was in Italy a restoration of the former local dynasties and powers, but the rule of primogeniture was not

TABLE 7.7. Sardinia (1800–1970)—second cousins: matrix of types of pedigrees

| Δ | K | | | | | Total | Mean frequency |
|---|---|---|---|---|---|---|---|
| | 0 | 1 | 2 | 3 | 4 | | |
| −2d | — | — | 5 : 388 | — | — | 388 | **388.00** |
| −d | — | 1 : 524<br>4 : 466<br>**495.0** | — | 7 : 579<br>13 : 452<br>**515.5** | — | 2021 | **505.25** |
| 0 | 0 : 663<br>**663** | — | 3 : 624<br>6 : 444<br>9 : 466<br>12 : 521<br>**513.8** | — | 15 : 728<br>**728** | 3446 | **574.33** |
| d | — | 2 : 551<br>8 : 560<br>**555.5** | — | 11 : 600<br>14 : 524<br>**562.0** | — | 2235 | **558.75** |
| 2d | — | — | 10 : **512** | — | — | 512 | **512.00** |
| Total | 663 | 2101 | 2955 | 2155 | 728 | 8602 | |
| Mean frequency | **663.00** | **525.25** | **492.50** | **538.25** | **728.00** | | **537.62** |

*Note.* Rows are categories of expected age difference, Δ; columns are male numbers among intermediate ancestors, *k*. Code of pedigree in italic type, frequency in roman type. Numbers in bold are the average frequencies of each cell of the matrix.

TABLE 7.8. Sardinia (1800–1970)—2½ cousins: matrix of types of pedigrees

| Δ | 0 | 1 | 2 | 3 | 4 | 5 | Total | Mean frequency |
|---|---|---|---|---|---|---|---|---|
| **Bride in the shorter branch** | | | | | | | | |
| $-h-2d$ | — | — | — | 42 : 6 | — | — | 6 | **6.00** |
| $-h-d$ | — | — | — | — | 43 : 12<br>46 : 3<br>**7.5** | — | 38 | **7.60** |
| $-h$ | — | 2 : 16<br>8 : 14<br>32 : 14<br>**14.7** | — | 11 : 6  38 : 6<br>14 : 8  41 : 14<br>35 : 13  44 : 18<br>**10.8** | — | 47 : 27<br>**27** | 136 | **13.60** |
| $-h+d$ | 0 : 32<br>**32** | — | 3 : 37  12 : 23<br>6 : 39  33 : 35<br>9 : 25  36 : 29<br>**31.3** | — | 15 : 26<br>39 : 29<br>45 : 21<br>**25.3** | — | 296 | **29.60** |
| $-h+2d$ | — | 1 : 38<br>4 : 40<br>**39.0** | — | 7 : 43<br>13 : 47<br>37 : 44<br>**44.7** | — | — | 212 | **42.40** |
| $-h+3d$ | — | — | 5 : 78 | — | — | — | 78 | **78.00** |
| Total | 32 | 122 | 289 | 205 | 91 | 27 | 766 | |
| Mean frequency | **32.00** | **24.40** | **28.90** | **20.50** | **18.20** | **27.00** | | **23.94** |

**Groom in the shorter branch**

| Groom in the shorter branch | | | | | | | Total | Mean frequency |
|---|---|---|---|---|---|---|---|---|
| h − 3d | — | — | *21* : 98 | — | — | — | 98 | **98.00** |
| h − 2d | — | *17* : 97<br>*20* : 100<br>**98.5** | — | *23* : 90<br>*29* : 82<br>*53* : 103<br>**91.7** | — | — | 472 | **94.40** |
| h − d | *16* : 68<br>**68** | — | *19* : 74  *28* : 62<br>*22* : 67  *49* : 67<br>*25* : 70  *52* : 61<br>**66.8** | — | *31* : 95<br>*55* : 76<br>*61* : 74<br>**81.7** | — | 714 | **71.40** |
| h | — | *18* : 48<br>*24* : 42<br>*48* : 39<br>**43.0** | — | *27* : 41  *54* : 43<br>*30* : 24  *57* : 45<br>*51* : 43  *60* : 37<br>**38.8** | — | *63* : 42<br>**42** | 404 | **40.40** |
| h + d | — | — | *26* : 23<br>*50* : 15<br>*56* : 16<br>**18.0** | — | *59* : 42<br>*62* : 19<br>**30.5** | — | 115 | **23.00** |
| h + 2d | — | — | — | *58* : **11** | — | — | 11 | **11.00** |
| Total | 68 | 326 | 553 | 519 | 306 | 42 | 1814 | |
| Mean frequency | **68.00** | **65.20** | **55.30** | **51.90** | **61.20** | **42.00** | | **56.69** |

*Note.* Rows are categories of expected age difference, $\Delta$; columns are male numbers among intermediate ancestors, $k$. Code of pedigree in italic type, frequency in roman type. Numbers in bold are the average frequencies of each cell of the matrix.

TABLE 7.9. Sardinia (1800–1970)—third cousins: matrix of types of pedigrees

| Δ | K | | | | | | | Total | Mean frequency |
|---|---|---|---|---|---|---|---|---|---|
| | 0 | 1 | 2 | 3 | 4 | 5 | 6 | | |
| −3d | — | — | — | 21 : 49 | — | — | — | 49 | **49.00** |
| −2d | — | — | 5 : 56<br>17 : 40<br>20 : 49<br>**48.3** | — | 23 : 56<br>29 : 46<br>53 : 48<br>**50.0** | — | — | 295 | **49.17** |
| −d | — | 1 : 76<br>4 : 72<br>16 : 58<br>**68.7** | — | 7 : 86   28 : 55<br>13 : 54   37 : 52<br>19 : 72   49 : 48<br>22 : 34   52 : 44<br>25 : 37<br>**53.5** | — | 31 : 96<br>55 : 75<br>61 : 52<br>**74.3** | — | 911 | **60.73** |

| | | | | | | | | Total | Mean frequency |
|---|---|---|---|---|---|---|---|---|---|
| **0** | 0 : 95 | — | 3 : 82  24 : 68<br>6 : 58  33 : 52<br>9 : 65  36 : 57<br>12 : 51  48 : 55<br>18 : 36 | — | 15 : 112  51 : 75<br>27 : 85  54 : 57<br>30 : 53  57 : 56<br>39 : 76  60 : 51<br>45 : 57 | — | 63 : 141 | 1382 | **69.10** |
| | **95** | | **58.2** | | **69.1** | | **141** | | |
| **d** | — | 2 : 51<br>8 : 90<br>32 : 71 | — | 11 : 80  41 : 56<br>14 : 60  44 : 41<br>26 : 48  50 : 59<br>35 : 92  56 : 53<br>38 : 49 | — | 47 : 116<br>59 : 83<br>62 : 67 | — | 1016 | **67.73** |
| | | **70.7** | | **59.8** | | **88.7** | | | |
| **2d** | — | — | 10 : 80<br>34 : 77<br>40 : 46 | — | 43 : 92<br>46 : 66<br>58 : 47 | — | — | 408 | **68.00** |
| | | | **67.7** | | **68.3** | | | | |
| **3d** | — | — | — | 42 : 44 | — | — | — | 44 | **44.00** |
| | | | | **44** | | | | | |
| **Total** | 95 | 418 | 872 | 1113 | 977 | 489 | 141 | 4105 | |
| **Mean frequency** | **95.00** | **69.67** | **58.13** | **55.65** | **65.13** | **81.50** | **141.00** | | **64.14** |

*Note.* Rows are categories of expected age difference, Δ; columns are male numbers among intermediate ancestors, *k*. Code of pedigree in italic type, frequency in roman type. Numbers in bold are the average frequencies of each cell of the matrix.

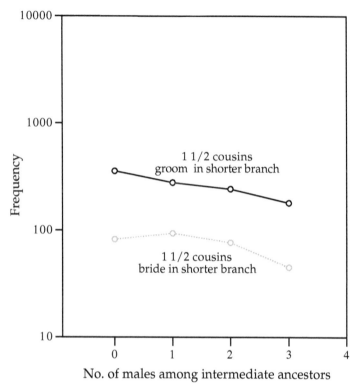

FIGURE 7.3. Sardinia 1850–1970. Average frequency of pedigrees as a function of the number of males among intermediate ancestors in first cousins once removed (1½).

reinstated and this may have favored a gradually increasing recourse to first-cousin marriages to avoid land fragmentation caused by the loss of primogeniture rules. The first-cousin marriage increase shows considerable parallelism in the various parts of Italy, including Sardinia and the other islands (see chapter 9, figure 9.12, from Moroni et al. [1992]). There is a slow increase at the beginning of the nineteenth century, which begins accelerating in the second half, reaching a peak in 1910–30, and then starts decreasing rapidly.

Another factor, not alternative to the avoidance of land fragmentation, but which can be added to it and may help explain the rapid increase at the end of the nineteenth century, is that the change of customs, favoring first-cousin matings and, in some parts of Italy (see

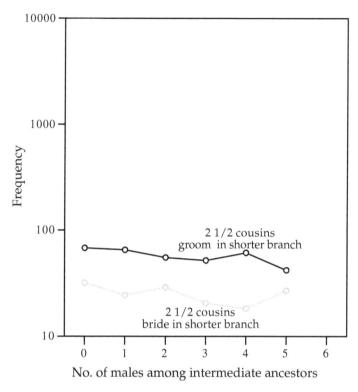

FIGURE 7.4. Sardinia 1850–1970. Average frequency of pedigrees as a function of the number of males among intermediate ancestors in second cousins once removed (2½).

section 7.3), even uncle–niece marriages, may have been accelerated by a partial loss of political influence of the Catholic Church that took place in the second part of the century. In 1860 the Kingdom of Savoy, later to become the Kingdom of Italy, occupied the territory of the pontifical state, which included practically all the central part of the Italian peninsula. In 1870 it took Rome by siege, which was the last part of the pontifical state still remaining under the pontiff's rule. The pontiff retired to the Vatican, which is today again a sovereign territory, though of very limited area, and was recognized as an independent state by Mussolini in the 1928 "Concordato" with the Holy See.

The decline of first-cousin marriages after the war might be seen,

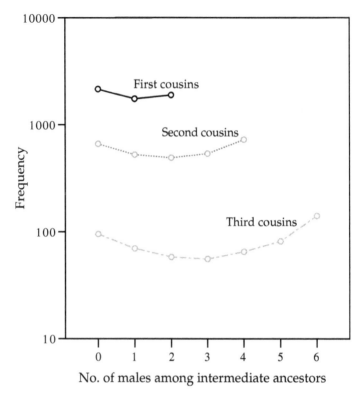

FIGURE. 7.5. Sardinia 1850–1970. Average frequency of pedigrees as a function of the number of males among intermediate ancestors in first, second, and third cousins.

at least in part, as a manifestation of the "breakdown of isolates" (Dahlberg 1948), connected with the increase in transportation and urbanization, and with the decline of agricultural work. Railways, however, did not increase considerably in Italy after the year 1900. Other modern means of transportation increased in a major way only in the second half of the nineteenth century. The role of transportation increase may then have been less important, directly, than the other two factors. Cousin marriage is usually infrequent in cities, and therefore urbanization is most likely a direct cause of a decrease in consanguineous marriages. The decrease during the twentieth century of the proportion of the Italian population engaged in agricultural activities contributed in part to urbanization, but also to the increase

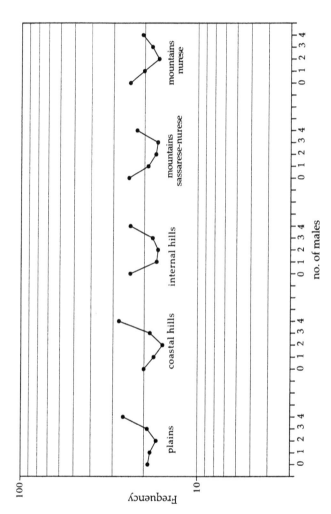

FIGURE 7.6. Sardinia 1850–1970. Second cousins. Average frequency of pedigrees as a function of the number of males among intermediate ancestors in five ecological areas.

TABLE 7.10. Sardinia (1850–1970)—second cousins: differential
migration in the five ecological areas

| | Number of males | | | | |
|---|---|---|---|---|---|
| | 0 | 4 | 0 + 4 | 0/(0 + 4) | 4/0 |
| Plains | 101 | 138 | 239 | 42.3 | 1.37 |
| Coastal hills | 105 | 145 | 250 | 42.0 | 1.38 |
| Internal hills | 348 | 350 | 698 | 49.9 | 1.00 |
| Mountains: Sassari-Nuoro | 61 | 55 | 116 | 52.6 | 0.90 |
| Mountains: Nuoro | 47 | 40 | 87 | 54.0 | 0.85 |
| | 662 | 728 | 1390 | | |

$$\chi^2_{[4]} = 9.89$$
$$P < 0.05$$

in geographic mobility due to relocation in the search of jobs in activities other than agriculture. Such an increase of geographic mobility must also decreases the overall chance of consanguineous
marriage.

Another factor that must have contributed to the change in consanguinity is the population increase, which was continuous throughout
the nineteenth century and started decreasing, eventually becoming
negative, during the twentieth century. In an expanding population
the proportion of consanguineous relatives increases, but with a delay

TABLE 7.11. Sardinia (1850–1970)—second cousins: differential migration
by type of economy

| | Number of males | | | | |
|---|---|---|---|---|---|
| | 0 | 4 | 0 + 4 | 0/(0 + 4) | 4/0 |
| Plains + coastal hills (farmers) | 206 | 283 | 489 | 42.1 | 1.37 |
| Mountains (shepherds) | 108 | 95 | 203 | 53.2 | 0.88 |
| | 314 | 378 | 692 | | |

$$\chi^2_{[1]} = 7.09$$
$$P < 0.01$$

of a number of generations, which depends on the consanguinity degree: it is earlier for first cousins and later and less pronounced, being more damped by migration, for less close cousins. The fact that second cousins are much less affected than first cousins by the change in time shows, however, that demographic change may not be the major factor.

The availability of a number of third-cousin pedigrees is useful for another type of analysis, which we have already done in Emilia (Section 4.3), where it was limited to third cousins, because they form the most informative set of marriages from this point of view: that of the effect of the sex of most remote ancestors and of those nearest to the consanguineous mates. We repeat it here on a wider scale for the third cousins of Sardinia, and will do it again in the next section for Sicily.

We chose for this purpose four subsets of third cousins, each of which has the same $\Delta$ (age difference between mates) and $k$ (number of males among intermediate ancetors) values: **A** is $\Delta = d, k = 3$; **B** is $\Delta = 0, k = 2$; **C** is $\Delta = 0, k = 4$; and **D** is $\Delta = -d, k = 3$. So, by carrying comparisons between the frequencies of the various types of mating only within each subset, we can eliminate the influence of age difference and sex of intermediate ancestors. Each subset is made of nine pedigrees, listed for the four subsets in tables 7.12 and 7.13. In table 7.12 we test the effect on the frequency of consanguineous marriages of the sex of the nearest intermediate ancestors (the parents of the consanguineous mates); in table 7.13 we test the effect of the sex of the most remote intermediate ancestors (the children of the common ancestors). In both tables we compare three categories of pedigrees: with two males (MM), one male–one female (MF), or two females (FF). The original frequencies of all the pedigrees used in these calculations are given in table 7.9.

The comparison of the sexes of nearest ancestors in table 7.12 shows that observed FF frequencies are always greater than expected, and MF frequencies are usually intermediate, or, in general, closer to the mean of the MM pedigrees. Significance tests were made with chi square, and two out of four are very highly significant. Only one out of four is significant with the more conservative ANOVA test (used additionally to account for a high variance within the MM, MF, and FF groups of frequencies). ANOVA is significant ($F_{[2,33]} = 3.37$,

$P = 0.046$), confirming the conclusion that the most frequent pair of nearest ancestors are sisters. We interpret this as evidence to confirm the conclusion of chapter 4 (section 4.3) that mothers of consanguineous mates are influential in determining marriages, even in third-cousin marriages, where they are three generations removed from the common ancestors.

A symmetric test, using the same subsets of pedigrees, allows us to compare the frequency of most remote ancestors, again subdivided into MM, MF, and FF (table 7.13). We would expect a decrease of the observed frequency of MM marriages compared with the expected, under a frequent application of the rule of primogeniture, decreasing the chance of more than one male child to marry; but we find, on the contrary, an increase. ANOVA confirms that the difference is very highly significant: $F_{[2,33]} = 40.40$, $P < 0.001$. As to the interpretation, a change with time of female migration at marriage was higher three generations ago and later diminished. It is also possible that this result is a consequence of social inequality: richer families may have more male children marrying; whether with consanguineous wives or not is irrelevant.

Two comparisons, those of nearest ancestors and those of the most remote ancestors, are highly correlated because of the structure of pedigrees and cannot be tested independently one from the other with this type of analysis. If it is correct that two females are more frequently nearest ancestors than other combinations, then the most remote ancestors will also tend to be two brothers. Hence, it is possible that only one of the two conclusions is true: (1) some mothers of consanguineous mates have favored alliances of their children, and (2) consanguineous mates who descend from two brothers have a slightly higher probability of mating. It is also possible that both hypotheses are true. One cannot choose among them on this evidence alone. It is reasonable to favor the hypothesis that females who are mothers of the consanguineous mates are more likely to play a part, because it would be difficult to explain otherwise the same phenomenon observed on the frequencies of first-cousin marriages. The hypothesis could be tested by a sociological analysis of the closeness with which family ties are maintained by relatives of either sex. Episodic observations suggest that this may be correct.

TABLE 7.12. Sardinia (1800–1970): effect of the sex of the parents of consanguineous mates (the nearest intermediate ancestors) on the frequency of third-cousin marriages grouped for the same $\Delta$ and $k$ values (from table 7.9)

| Group | Two males | One male–<br>one female | Two females | Total |
|---|---|---|---|---|
| A: $\Delta = d; k = 3$ | | | | |
| Pedigree codes | 50, 56 | 26, 35, 38, 41, 44 | 11,14 | |
| Observed no. of marriages | 112 | 286 | 140 | 538 |
| Expected no. of marriages | 119.6 | 298.9 | 119.6 | 538.1 |
| $\chi^2_{[2]}$ | 0.48 | 0.56 | 3.48 | 4.52 |
| | | | | $P = 0.104$ |
| B: $\Delta = 0; k = 2$ | | | | |
| Pedigree codes | 48 | 18, 24, 33, 36 | 3, 6, 9, 12 | |
| Observed no. of marriages | 55 | 213 | 256 | 524 |
| Expected no. of marriages | 58.2 | 232.9 | 232.9 | 524.0 |
| $\chi^2_{[2]}$ | 0.17 | 1.70 | 2.29 | 4.16 |
| | | | | $P = 0.124$ |
| C: $\Delta = 0; k = 4$ | | | | |
| Pedigree codes | 51, 54, 57, 60 | 27, 30, 39, 45 | 15 | |
| Observed no. of marriages | 239 | 271 | 112 | 622 |
| Expected no. of marriages | 276.4 | 276.4 | 69.1 | 621.9 |
| $\chi^2_{[2]}$ | 5.06 | 0.10 | 26.63 | 31.79 |
| | | | | $P < 0.0001$ |
| D: $\Delta = -d; k = 3$ | | | | |
| Pedigree codes | 49, 52 | 19, 22, 25, 28, 37 | 7, 13 | |
| Observed no. of marriages | 92 | 250 | 140 | 482 |
| Expected no. of marriages | 107.1 | 267.8 | 107.1 | 482.0 |
| $\chi^2_{[2]}$ | 2.13 | 1.18 | 10.11 | 13.42 |
| | | | | $P = 0.001$ |
| Total mean frequency of marriages ± standard error | 55.33 ± 2.91 | 56.67 ± 3.86 | 72.00 ± 6.58 | 60.17 ± 2.81 |

TABLE 7.13. Sardinia (1800–1970): effect of the sex of remotest intermediate ancestors on the frequency of third-cousin marriages grouped for the same $\Delta$ and $k$ values (from table 7.9)

| Group | Two males | One male– one female | Two females | Total |
|---|---|---|---|---|
| A: $\Delta = d; k = 3$ | | | | |
| Pedigree codes | 11, 35 | 14, 26, 38, 41, 50 | 44, 56 | |
| Observed no. of marriages | 172 | 272 | 94 | 538 |
| Expected no. of marriages | 119.6 | 298.9 | 119.6 | 538.1 |
| $\chi^2_{[2]}$ | 22.96 | 2.42 | 5.48 | 30.86 |
| | | | | $P < 0.0001$ |
| B: $\Delta = 0; k = 2$ | | | | |
| Pedigree codes | 3 | 6, 9, 18, 33 | 12, 24, 36, 48 | |
| Observed no. of marriages | 82 | 211 | 231 | 524 |
| Expected no. of marriages | 58.2 | 232.9 | 232.9 | 524.0 |
| $\chi^2_{[2]}$ | 9.73 | 2.06 | 0.01 | 11.80 |
| | | | | $P = 0.003$ |
| C: $\Delta = 0; k = 4$ | | | | |
| Pedigree codes | 15, 27, 39, 51 | 30, 45, 54, 57 | 60 | |
| Observed no. of marriages | 348 | 223 | 51 | 622 |
| Expected no. of marriages | 276.4 | 276.4 | 69.1 | 621.9 |
| $\chi^2_{[2]}$ | 18.55 | 10.32 | 4.74 | 33.61 |
| | | | | $P < 0.0001$ |
| D: $\Delta = -d; k = 3$ | | | | |
| Pedigree codes | 7, 19 | 13, 22, 25, 37, 49 | 28, 52 | |
| Observed no. of marriages | 158 | 225 | 99 | 482 |
| Expected no. of marriages | 107.1 | 267.8 | 107.1 | 482.0 |
| $\chi^2_{[2]}$ | 24.19 | 6.84 | 0.61 | 31.64 |
| | | | | $P < 0.0001$ |
| Total mean frequency of marriages ± standard error | 84.44 ± 4.01 | 51.72 ± 2.02 | 52.77 ± 2.59 | 60.17 ± 2.81 |

## 7.3 SICILY

Sicily has an area slightly larger than Sardinia, but a much larger population (almost 5 million). It has a very slightly greater population density than the whole of Italy, from which it is separated by the very narrow Straits of Messina. It is mountainous, with the highest mountain, Etna, an active volcano, reaching 3263 meters. It has been inhabited since Paleolithic times and had active Neolithic settlers, who shared with those of Calabria, the nearest region of the Italian peninsula, access to the island of Lipari in the Aeolian Islands, particularly rich in obsidian. Its autochthonous inhabitants were called Sicani and are believed to have been forced to the western part by the Siculi, who came from the peninsula and were of "italic" origin, speaking Indo-European languages, and are believed to have arrived to Italy from the northeast, around the tenth century B.C.

Sicily was colonized in the eighth century B.C. by Phoenicians, and their very successful descendants, the Carthaginians, who settled first Tunisia and then also Spain. Greeks formed many colonies, especially in the eastern part, and fought the Carthaginians successfully, but were subdued by the Romans beginning in 241 B.C. In the fifth century A.D. toward the end of the western Roman Empire, Sicily (and also Sardinia) were invaded by Vandals, who later left for northern Africa, and by Ostrogoths, but in the sixth century A.D. Sicily became part of the eastern Roman Empire. Arab domination began in A.D. 827 and lasted 250 years, to be followed by the Normans. After a century the emperors of Germany, then the French for a short time, and the Spanish in 1409 took over. The Spanish soon formed a joint kingdom of Sicily and the southern part of the Italian peninsula, formerly called the Kingdom of Naples and later the Kingdom of the Two Sicilies. This union continued, after short occupations by other powers, in the eighteenth and the nineteenth centuries, until 1860, when Garibaldi, the leading Italian patriot, occupied Sicily and defeated the Kingdom of Naples. Soon thereafter, the south of Italy was united with the northern part of Italy, which had been conquered by the king of Savoy.

Consanguinity data in Sicily were collected by A. Moroni and his collaborators (1972) in a sample of dioceses, about one-fifth of all

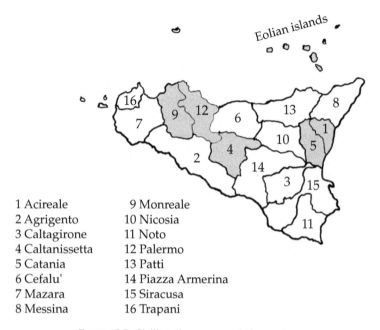

1 Acireale                 9 Monreale
2 Agrigento                10 Nicosia
3 Caltagirone              11 Noto
4 Caltanissetta            12 Palermo
5 Catania                  13 Patti
6 Cefalu'                  14 Piazza Armerina
7 Mazara                   15 Siracusa
8 Messina                  16 Trapani

FIGURE 7.7. Sicilian dioceses: sampled areas in gray.

those existing in Sicily (dioceses sampled are shown in figure 7.7). Table 7.14 indicates the absolute and relative frequencies (‰) of consanguineous marriages by degree since 1685, cumulatively in this sample of dioceses. The total number of marriages increases proportionately to the total population, which approximately doubled from 1850 to 1950, in spite of a population loss in the period 1890–1900, most probably because of much emigration to America and population decrease during World War I. It is more difficult to give estimates for periods before 1850 because of incomplete records.

Third cousins show only a small decrease of relative frequency in the period in which they are recorded, and second cousins are constant through the eighteenth century, showing only a slow decrease during the twentieth century. First cousins, uncle–niece, and to a lesser extent 1½ cousins show a major increase, of almost one order of magnitude, between 1850 and the period 1918–1929. After this time interval relative frequencies start decreasing toward almost "normal" values. "12"-type marriages (uncle–niece and aunt–nephew) show a remarkable parallelism with first cousins.

Sicily is the only region of Italy, except for the neighboring Calabria (and, in particular, its more southern part, close to Sicily across the Straits of Messina), in which "12"-type marriages reach a remarkably high frequency, though only for the short period of little over a century, beginning in 1850. It is of interest to study the geographic distribution of "12" marriages in more detail. This is possible in the Vatican data. Table 7.15 shows them grouped by the nine provinces. Two provinces, Enna (EN) and Ragusa (RG), are recent, having been carved from the provinces of Caltanissetta (CL) and Syracuse (SR), respectively, and explain the absence of information until 1930. Unlike our data from dioceses, the Vatican data are complete for the whole island, but are limited to closest types of cousins and only to the latest period (1910–1964). The highest relative frequencies of uncle–niece marriages (aunt–nephew is only a relatively small fraction of the "12" total) are observed in Agrigento, Caltanissetta, and Syracuse, that is, in the south and center. First cousins (given at the bottom of table 7.15) show less geographic variation, also, but not necessarily only because they occur in larger numbers and are therefore less subject to statistical variation.

The diocese data allow us to count pedigree types and carry out the same analyses that were made in Emilia (section 4.3) and Sardinia (section 7.2). In table 7.17 first cousins show, as usual, a higher frequency of marriages when parents of consanguineous mates are sisters, and the age effect shows the maximum for an age difference of 0, as in all other types of even cousins (tables 7.19 and 7.21). The effect of numbers of males among intermediate ancestors is similar to that observed in Emilia—see also the summary figure of results in even cousins (figure 7.8), where the phenomenon is more clearly observed, and therefore speaks in favor of a regular patrilocal custom, whereby women move to the husband's house at marriage. Uneven cousins (tables 7.16, 7.18, and 7.20) show, as in the other areas, a strong effect of age difference in the expected direction.

The analysis of nearest intermediate ancestors (table 7.22) shows that, as in Sardinia, the most frequent pedigree on average is of FF type, and this is true in all four subsets tested. One of the four sets is significant at $P < 0.0001$. The combined ANOVA, however, is not significant ($F_{[2,33]} = 1.24$, $P = 0.303$). The analysis of most remote ancestors (table 7.23) gives complementary conclusions, as observed in Sardinia, and the combined ANOVA is significant ($F_{[2,33]}$

TABLE 7.14. Frequencies of consanguineous marriages in Sicily

| Years | Consanguineous marriages | | | | | | | Total marriages |
|---|---|---|---|---|---|---|---|---|
| | 12 | 22 | 23 | 33 | 34 | 44 | Multiple | |
| 1685–1849 | 0 | 591 | 761 | 1,982 | 1,252 | 1,603 | 120 | 140,624 |
| 1850–59 | 14 | 135 | 189 | 438 | 310 | 361 | 82 | 29,592 |
| 1860–69 | 22 | 228 | 159 | 331 | 195 | 209 | 45 | 30,406 |
| 1870–79 | 46 | 690 | 283 | 548 | 223 | 256 | 109 | 35,212 |
| 1880–89 | 75 | 1,118 | 375 | 699 | 326 | 315 | 140 | 40,142 |
| 1890–99 | 103 | 1,418 | 451 | 758 | 308 | 322 | 175 | 39,284 |
| 1900–09 | 139 | 1,905 | 525 | 813 | 305 | 306 | 201 | 47,437 |
| 1910–17 | 108 | 1,459 | 326 | 505 | 179 | 196 | 159 | 30,255 |
| 1918–29 | 270 | 3,641 | 571 | 951 | — | — | 184 | 63,775 |
| 1930–39 | 190 | 2,670 | 430 | 823 | — | — | 135 | 53,256 |
| 1940–49 | 168 | 2,666 | 543 | 779 | — | — | 104 | 57,728 |
| 1950–59 | 117 | 2,742 | 558 | 875 | — | — | 99 | 65,266 |
| 1960–69 | 57 | 1,764 | 402 | 706 | — | — | 80 | 54,704 |

Consanguineous marriages ‰

| Years | 12 | 22 | 23 | 33 | 34 | 44 | Multiple | Total marriages % | α‰ |
|---|---|---|---|---|---|---|---|---|---|
| 1685–1849 | 0 | 4.20 | 5.41 | 14.09 | 8.90 | 11.40 | 0.85 | 4.49 | 0.80 |
| 1850–59 | 0.47 | 4.56 | 6.39 | 14.80 | 10.48 | 12.20 | 2.77 | 5.17 | 0.98 |
| 1860–69 | 0.72 | 7.50 | 5.23 | 10.89 | 6.41 | 6.87 | 1.48 | 3.91 | 1.04 |
| 1870–79 | 1.31 | 19.60 | 8.04 | 15.56 | 6.33 | 7.27 | 3.10 | 6.12 | 2.08 |
| 1880–89 | 1.87 | 27.85 | 9.34 | 17.41 | 8.12 | 7.85 | 3.49 | 7.59 | 2.79 |
| 1890–99 | 2.62 | 36.10 | 11.48 | 19.30 | 7.84 | 8.20 | 4.45 | 9.00 | 3.56 |
| 1900–09 | 2.93 | 40.16 | 11.07 | 17.14 | 6.43 | 6.45 | 4.24 | 8.84 | 3.80 |
| 1910–17 | 3.57 | 48.22 | 10.78 | 16.69 | 5.92 | 6.48 | 5.26 | 9.69 | 4.46 |
| 1918–29 | 4.23 | 57.09 | 8.95 | 14.91 | — | — | 2.89 | 8.81 | 4.87 |
| 1930–39 | 3.57 | 50.14 | 8.07 | 15.45 | — | — | 2.53 | 7.98 | 4.30 |
| 1940–49 | 2.91 | 46.18 | 9.41 | 13.49 | — | — | 1.80 | 7.38 | 3.92 |
| 1950–59 | 1.79 | 42.01 | 8.55 | 13.41 | — | — | 1.52 | 6.73 | 3.45 |
| 1960–69 | 1.04 | 32.25 | 7.35 | 12.91 | — | — | 1.46 | 5.50 | 2.71 |

TABLE 7.15. Frequencies (‰) of uncle–niece and aunt–nephew marriages in the nine provinces of Sicily

| Years | AG | CL | EN | CT | ME | PA | SR | RG | TP |
|-------|------|------|------|------|------|------|------|------|------|
| 1910–14 | 5.18 | 7.83 | — | 3.03 | 2.37 | 2.36 | 5.66 | — | 0.80 |
| 1915–19 | 5.71 | 7.74 | — | 4.41 | 3.18 | 2.10 | 6.21 | — | 1.19 |
| 1920–24 | 7.93 | 7.24 | — | 4.52 | 3.03 | 3.84 | 7.15 | — | 1.85 |
| 1925–29 | 6.89 | 7.27 | — | 4.36 | 2.12 | 2.77 | 6.12 | — | 1.18 |
| 1930–34 | 5.38 | 7.24 | 5.72 | 3.32 | 2.69 | 2.07 | 6.66 | 6.12 | 1.21 |
| 1935–39 | 6.54 | 4.49 | 5.29 | 3.84 | 2.17 | 2.48 | 4.92 | 4.55 | 0.79 |
| 1940–44 | 4.85 | 3.96 | 5.56 | 1.76 | 1.78 | 1.80 | 3.55 | 3.88 | 0.39 |
| 1945–49 | 7.33 | 3.95 | 5.06 | 2.88 | 2.05 | 2.13 | 4.93 | 4.24 | 0.82 |
| 1950–54 | 5.03 | 4.34 | 5.01 | 1.56 | 1.84 | 1.36 | 3.30 | 3.16 | 0.77 |
| 1955–59 | 3.14 | 2.61 | 3.54 | 1.38 | 1.20 | 0.65 | 1.91 | 2.01 | 0.42 |
| 1960–64 | 2.32 | 2.05 | 4.22 | 0.68 | 0.83 | 0.56 | 1.32 | 1.31 | 0.23 |
| **Average** | **5.42** | **5.33** | **4.87** | **2.81** | **2.06** | **1.93** | **4.93** | **3.55** | **0.86** |
| First cousins | 53.14 | 43.55 | 45.57 | 37.56 | 51.07 | 46.03 | 42.62 | 37.94 | 44.79 |

*Source.* Data from Vatican archives.

*Note.* AG, Agrigento; CL, Caltanissetta; EN, Enna; CT, Catania; ME, Messina, PA, Palermo; SR, Syracuse; RG, Ragusa; TP, Trapani.

TABLE 7.16. Sicily (1800–1970)—uncle–niece and aunt–nephew: matrix of types of pedigrees

| $\Delta$ | K | | | Total | Mean frequency |
|---|---|---|---|---|---|
| | 0 | 1 | 2 | | |
| **Aunt–nephew** | | | | | |
| $-h$ | — | *2 :* **30** | — | 30 | **30.00** |
| $-h + d\ (-w)$ | *0 :* **86** | — | — | 86 | **86.00** |
| **Uncle–niece** | | | | | |
| $h - d\ (w)$ | — | *1 :* **714** | — | 714 | 714.00 |
| $h$ | — | — | *3 :* **310** | 310 | **310.00** |
| Total | 86 | 744 | 310 | 1150 | |
| Mean frequency | **86.00** | **372.00** | **310.00** | | **287.5** |

*Note.* Rows are categories of expected age difference, $\Delta$; columns are male numbers among intermediate ancestors, $k$. Code of pedigree in italic type, frequency in roman type. Numbers in bold are the average frequencies of each cell of the matrix.

TABLE 7.17. Sicily (1800–1970)—first cousins: matrix of types of pedigrees

| | K | | | | Mean |
| Δ | 0 | 1 | 2 | Total | frequency |
|---|---|---|---|---|---|
| −d | — | *1* : **3,369** | — | 3,369 | **3,369.00** |
| 0 | *0* : **5,700** | — | *3* : **4,384** | 10,084 | **5,042.00** |
| d | — | *2* : **4,908** | — | 4,908 | **4,908.00** |
| Total | 5,700 | 8,277 | 4,384 | 18,361 | |
| Mean frequency | **5,700.00** | **4,138.50** | **4,384.00** | | **4,590.25** |

*Note.* Rows are categories of expected age difference, Δ; columns are male numbers among intermediate ancestors, $k$. Code of pedigree in italic type, frequency in roman type. Numbers in bold are the average frequencies of each cell of the matrix.

TABLE 7.18. Sicily (1800–1970)—1½ cousins: matrix of types of pedigrees

| | K | | | | | Mean |
| Δ | 0 | 1 | 2 | 3 | Total | frequency |
|---|---|---|---|---|---|---|
| **Bride in the shorter branch** | | | | | | |
| −h − d | — | — | *10* : **45** | — | 45 | **45.00** |
| | | *2* : 69 | | | | |
| −h | — | *8* : 61 | — | *11* : 144 | 274 | **91.33** |
| | | **65.0** | | **144** | | |
| | | | *3* : 180 | | | |
| −h + d | *0* : 170 | — | *9* : 143 | — | 493 | **164.33** |
| | **170** | | **161.5** | | | |
| −h + 2d | — | *1* : **245** | — | | 245 | **245.00** |
| Total | 170 | 375 | 368 | 144 | 1057 | |
| Mean frequency | **170.00** | **125.00** | **122.67** | **144.00** | | **132.12** |
| **Groom in the shorter branch** | | | | | | |
| h − 2d | — | *5* : **614** | — | — | 614 | **614.00** |
| | | | *7* : 605 | | | |
| h − d | *4* : 539 | — | *13* : 413 | — | 1557 | **519.00** |
| | **539** | | **509.0** | | | |
| | | *6* : 261 | | | | |
| h | — | *12* : 275 | — | *15* : 417 | 953 | **317.67** |
| | | **268.0** | | **417** | | |
| h + d | — | — | *14* : **159** | — | 159 | **159.00** |
| Total | 539 | 1150 | 1177 | 417 | 3283 | |
| Mean frequency | **539.00** | **383.33** | **392.33** | **417.00** | | **410.37** |

*Note.* Rows are categories of expected age difference, Δ; columns are male numbers among intermediate ancestors, $k$. Code of pedigree in italic type, frequency in roman type. Numbers in bold are the average frequencies of each cell of the matrix.

TABLE 7.19. Sicily (1800–1970)—second cousins: matrix of types of pedigrees

| Δ | K | | | | | Total | Mean frequency |
|---|---|---|---|---|---|---|---|
| | 0 | 1 | 2 | 3 | 4 | | |
| −2d | — | — | *5 : 290* | — | — | 290 | **290.00** |
| −d | — | *1 : 391*<br>*4 : 284*<br>**337.5** | — | *7 : 632*<br>*13 : 382*<br>**507.0** | — | 1689 | **422.25** |
| 0 | *0 : 570*<br>**570** | — | *3 : 659*<br>*6 : 370*<br>*9 : 363*<br>*12 : 438*<br>**437.5** | — | *15 : 1047*<br>**1047** | 3447 | **574.50** |
| d | — | *2 : 425*<br>*8 : 392*<br>**408.5** | — | *11 : 704*<br>*14 : 377*<br>**540.5** | — | 1898 | **474.50** |
| 2d | — | — | *10 : 331* | — | — | 331 | **331.00** |
| Total | 570 | 1492 | 2451 | 2095 | 1047 | 7655 | |
| Mean frequency | **570.00** | **373.00** | **408.50** | **523.75** | **1047.00** | | **478.44** |

*Note.* Rows are categories of expected age difference, Δ; columns are male numbers among intermediate ancestors, $k$. Code of pedigree in italic type, frequency in roman type. Numbers in bold are the average frequencies of each cell of the matrix.

$= 19.78$, $P < 0.0001$). Again, the two analyses are not independent, because the two pedigrees that have the two nearest female ancestors have also the two remotest brothers. We prefer to conclude that the effect is largely due to the two female nearest ancestors, but it is also possible that at least part of the effect is due to the two most remote ancestors, who are brothers.

## 7.4 AEOLIAN ISLANDS

The Aeolian Islands are in the Tyrrhenian Sea, north of Messina, eastern Sicily. The global surface is of 88 km$^2$. Seven of the islands are of sufficient size to be usually inhabited. The biggest of them, Lipari, has close to 75% of all inhabitants; the next in size, Salina, has less than 20%. The other five islands have from 100 to 400 inhabitants each.

A study of the history, demography, and consanguinity was carried out by Moroni (1967). The islands are all of volcanic origin, and were visited regularly and in part settled, especially Lipari, in the Neolithic period because of the abundance of obsidian. Lipari was the source of obsidian found in considerable abundance in Calabria and Sicily. It was later a normal stop in Cretan and Mycenean navigation in the second millennium B.C. Violent eruptions took place in the ninth century B.C. In the seventh century B.C. Lipari was settled by Greek colonists. Roman domination began in year 217 B.C., but the settlement was largely interrupted after the fall of the Roman empire. With the Normans in the eleventh century the islands began a period of prosperity, which ended in 1554 when almost all the inhabitants were enslaved and taken to Turkey by a famous pirate. Those who were not enslaved or were able to free themselves and new immigrants from all parts of south Italy reconstituted the population of the Aeolian islands. Since Norman times and until Italian unity the island was governed by the bishop of Lipari. The population peaked at about 20,000 inhabitants at the end of nineteenth and beginning of the twentieth centuries, and is now around 11,000 permanent residents, with numerous additions of tourists in the summer.

The average inbreeding coefficients in the various islands and in the ensemble, their variation in time, and their proportions due to the

TABLE 7.20. Sicily (1800–1970)—2½ cousins: matrix of types of pedigrees

| Δ | | | | K | | | Total | Mean frequency |
|---|---|---|---|---|---|---|---|---|
| | 0 | 1 | 2 | 3 | 4 | 5 | | |
| **Bride in the shorter branch** | | | | | | | | |
| −h − 2d | — | — | — | 42 : 2 | — | — | 2 | 2.00 |
| −h − d | — | — | 10 : 4<br>34 : 4<br>40 : 1<br>**3.00** | — | 43 : 14<br>46 : 15<br>**14.50** | — | 38 | 7.60 |
| −h | — | 2 : 9<br>8 : 9<br>32 : 6<br>**8.00** | — | 11 : 25  38 : 10<br>14 : 14  41 : 10<br>35 : 9  44 : 8<br>**12.67** | — | 47 : 38<br>**38** | 138 | 13.80 |
| −h + d | 0 : 14<br>**14** | — | 3 : 27  12 : 16<br>6 : 17  33 : 10<br>9 : 16  36 : 13<br>**16.50** | — | 15 : 45<br>39 : 22<br>45 : 14<br>**27.00** | — | 194 | 19.40 |
| −h + 2d | — | 1 : 21<br>4 : 31<br>**26.00** | — | 7 : 28<br>13 : 29<br>37 : 21<br>**26.00** | — | — | 130 | 26.00 |
| −h + 3d | — | — | 5 : 35<br>**35** | — | — | — | 35 | 35.00 |
| Total | 14 | 76 | 143 | 156 | 110 | 38 | 537 | |
| Mean frequency | **14.00** | **15.40** | **14.30** | **15.60** | **22.00** | **38.00** | | 16.78 |

**Groom in the shorter branch**

| | | | | | | | | |
|---|---|---|---|---|---|---|---|---|
| *h − 3d* | — | — | *21 : 59* | — | — | — | 59 | **59.00** |
| *h − 2d* | — | *17 : 57*<br>*20 : 39*<br>**48.00** | — | *23 : 68*<br>*29 : 35*<br>*53 : 47*<br>**50.00** | — | — | 246 | **49.20** |
| *h − d* | *16 : 43*<br>**43** | — | *19 : 42  28 : 25*<br>*22 : 41  49 : 32*<br>*25 : 31  52 : 39*<br>**35.00** | — | *31 : 102*<br>*55 : 46*<br>*61 : 39*<br>**62.33** | — | 440 | **44.00** |
| *h* | — | *18 : 22*<br>*24 : 24*<br>*48 : 18*<br>**21.33** | — | *27 : 40  54 : 26*<br>*30 : 20  57 : 29*<br>*51 : 37  60 : 26*<br>**29.67** | — | *63 : 87*<br>**87** | 329 | **32.90** |
| *h + d* | — | — | *26 : 10*<br>*50 : 11*<br>*56 : 12*<br>**11.00** | — | *59 : 28*<br>*62 : 12*<br>**20.00** | — | 73 | **14.60** |
| *h + 2d* | — | — | — | *58 : 10*<br>**10** | — | — | 10 | **10.00** |
| Total | 43 | 160 | 302 | 338 | 227 | 87 | 1157 | |
| Mean frequency | **43.00** | **32.00** | **30.20** | **33.80** | **45.40** | **87.00** | | **36.16** |

*Note.* Rows are categories of expected age difference, Δ; columns are male numbers among intermediate ancestors, *k*. Code of pedigree in italic type, frequency in roman type. Numbers in bold are the average frequencies of each cell of the matrix.

TABLE 7.21. Sicily (1800–1970)—third cousins: matrix of types of pedigrees

| Δ | K | | | | | | | Total | Mean frequency |
|---|---|---|---|---|---|---|---|---|---|
| | 0 | 1 | 2 | 3 | 4 | 5 | 6 | | |
| −3d | — | — | — | 21 : **15** | — | — | — | 15 | **15.00** |
| −2d | — | — | 5 : 14<br>17 : 18<br>20 : 10<br>**14.00** | — | 23 : 33<br>29 : 24<br>53 : 23<br>**26.67** | — | — | 122 | **20.33** |
| −d | — | 1 : 26<br>4 : 15<br>16 : 12 | — | 7 : 40  28 : 24<br>13 : 13  37 : 15<br>19 : 25  49 : 10<br>22 : 21  52 : 25<br>25 : 14 | — | 31 : 96<br>55 : 46<br>61 : 32 | — | 414 | **27.60** |
| | | **17.67** | | **20.77** | | **58.00** | | | |

| | | | | | | | | | |
|---|---|---|---|---|---|---|---|---|---|
| **0** | *0* : 38 **38** | — | *3* : 32  *6* : 21  *9* : 11  *12* : 15  *18* : 20  *24* : 17  *33* : 24  *36* : 18  *48* : 16 **19.33** | — | *15* : 85  *27* : 38  *30* : 18  *39* : 38  *45* : 17  *51* : 35  *54* : 18  *57* : 20  *60* : 33 **33.55** | — | *63* : 149 **149** | 663 | 33.15 |
| *d* | — | *2* : 20  *8* : 22  *32* : 22 **21.33** | — | *11* : 43  *14* : 17  *26* : 21  *35* : 42  *38* : 30  *41* : 20  *44* : 15  *50* : 21  *56* : 16 **25.00** | — | *47* : 119  *59* : 40  *62* : 27 **62.00** | — | 475 | 31.67 |
| *2d* | — | — | *10* : 21  *34* : 29  *40* : 22 **24.00** | — | *43* : 41  *46* : 19  *58* : 19 **26.33** | — | — | 151 | 25.17 |
| *3d* | — | — | — | *42* : 18 **18** | — | — | — | 18 | 18.00 |
| Total | 38 | 117 | 288 | 445 | 461 | 360 | 149 | 1858 | |
| Mean frequency | **38.00** | **19.50** | **19.20** | **22.25** | **30.73** | **60.00** | **149.00** | | **29.03** |

*Note.* Rows are categories of expected age difference, Δ; columns are male numbers among intermediate ancestors, *k*. Code of pedigree in italic type, frequency in roman type. Numbers in bold are the average frequencies of each cell of the matrix.

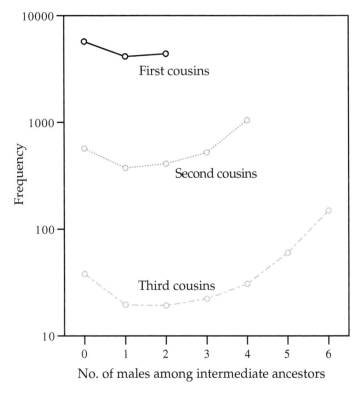

FIGURE 7.8. Sicily 1800–1970. Average frequency of pedigrees as a function of the number of males among intermediate ancestors in first, second, and third cousins.

various degrees of cousin marriages are shown in figure 7.9. The average inbreeding coefficient varied between 1 and 2‰ in the seventeenth century and started increasing slowly at the beginning of the eighteenth, reaching a sudden peak of 6‰ in the period 1800–1809, due to an increase of first cousins. Data are not available between 1844 and 1865; from then and until 1929 the number of first cousins remains very high and the inbreeding coefficient reaches almost 12‰ in 1910–1917, but then starts going down. At peak time, the largest island, Lipari, has average inbreeding around 0.8%, the next largest, Salina, just above 1%. The highest values, near or above 2%, are in Panarea (number of inhabitants varying between 236 and 790 in the

TABLE 7.22. Sicily (1800–1970): effect of the sex of the parents of consanguineous mates (the nearest intermediate ancestors) on the frequency of third-cousin marriages grouped for the same $\Delta$ and $k$ values (from table 7.21)

| Group | Two males | One male–<br>one female | Two females | Total |
|---|---|---|---|---|
| A: $\Delta = d$; $k = 3$ | | | | |
| Pedigree codes | 50, 56 | 26, 35, 38, 41, 44 | 11, 14 | |
| Observed no. of marriages | 37 | 128 | 60 | 225 |
| Expected no. of marriages | 50.0 | 125 | 50.0 | 225.0 |
| $\chi^2_{[2]}$ | 3.38 | 0.07 | 2.00 | 5.45 |
| | | | | $P = 0.065$ |
| B: $\Delta = 0$; $k = 2$ | | | | |
| Pedigree codes | 48 | 18, 24, 33, 36 | 3, 6, 9, 12 | |
| Observed no. of marriages | 16 | 79 | 79 | 174 |
| Expected no. of marriages | 19.3 | 77.3 | 77.3 | 173.9 |
| $\chi^2_{[2]}$ | 0.56 | 0.04 | 0.04 | 0.64 |
| | | | | $P = 0.724$ |
| C: $\Delta = 0$; $k = 4$ | | | | |
| Pedigree codes | 51, 54, 57, 60 | 27, 30, 39, 45 | 15 | |
| Observed no. of marriages | 106 | 111 | 85 | 302 |
| Expected no. of marriages | 134.2 | 134.2 | 33.6 | 301.9 |
| $\chi^2_{[2]}$ | 5.93 | 4.01 | 78.63 | 88.57 |
| | | | | $P < 0.0001$ |
| D: $\Delta = -d$; $k = 3$ | | | | |
| Pedigree codes | 49, 52 | 19, 22, 25, 28, 37 | 7, 13 | |
| Observed no. of marriages | 35 | 99 | 53 | 187 |
| Expected no. of marriages | 41.6 | 103.9 | 41.6 | 187.1 |
| $\chi^2_{[2]}$ | 1.05 | 0.23 | 3.12 | 4.40 |
| | | | | $P = 0.110$ |
| Total mean frequency of marriages ± standard error | 21.56 ± 2.72 | 23.17 ± 2.00 | 30.78 ± 7.83 | 24.67 ± 2.29 |

TABLE 7.23. Sicily (1800–1970): effect of the sex of remotest intermediate ancestors on the frequency of third-cousin marriages grouped for the same $\Delta$ and $k$ values (from table 7.21)

| Group | Two males | One male–one female | Two females | Total |
|---|---|---|---|---|
| A: $\Delta = d; k = 3$ | | | | |
| Pedigree codes | 11, 35 | 14, 26, 38, 41, 50 | 44, 56 | |
| Observed no. of marriages | 85 | 109 | 31 | 225 |
| Expected no. of marriages | 50.0 | 125.0 | 50.0 | 225.0 |
| $\chi^2_{[2]}$ | 24.50 | 2.05 | 7.22 | 33.77 |
| | | | | $P < 0.0001$ |
| B: $\Delta = 0; k = 2$ | | | | |
| Pedigree codes | 3 | 6, 9, 18, 33 | 12, 24, 36, 48 | |
| Observed no. of marriages | 32 | 76 | 66 | 174 |
| Expected no. of marriages | 19.33 | 77.33 | 77.33 | 174.0 |
| $\chi^2_{[2]}$ | 8.30 | 0.02 | 1.66 | 9.98 |
| | | | | $P = 0.007$ |
| C: $\Delta = 0; k = 4$ | | | | |
| Pedigree codes | 15, 27, 39, 51 | 30, 45, 54, 57 | 60 | |
| Observed no. of marrieages | 196 | 73 | 33 | 302 |
| Expected no. of marriages | 134.2 | 134.2 | 33.55 | 301.9 |
| $\chi^2_{[2]}$ | 28.46 | 27.91 | 0.01 | 56.38 |
| | | | | $P < 0.0001$ |
| D: $\Delta = -d; k = 3$ | | | | |
| Pedigree codes | 7, 19 | 13, 22, 25, 37, 49 | 28, 52 | |
| Observed no. of marriages | 65 | 73 | 49 | 187 |
| Expected no. of marriages | 41.6 | 103.9 | 41.6 | 187.1 |
| $\chi^2_{[2]}$ | 13.16 | 9.19 | 1.32 | 23.67 |
| | | | | $P < 0.0001$ |
| Total mean frequency of marriages ± standard error | 42.00 ± 5.68 | 18.38 ± 1.13 | 19.89 ± 2.05 | 24.67 ± 2.29 |

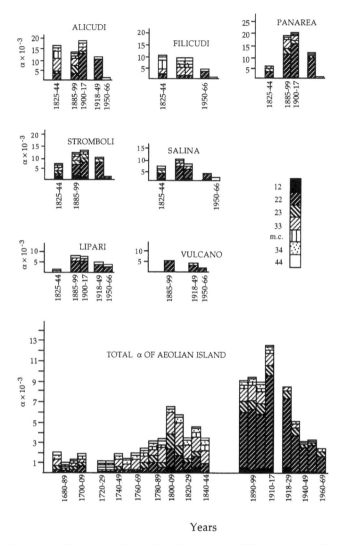

FIGURE 7.9. Variation in time of the inbreeding coefficient in the Aeolian Islands.

period 1825–1971) and Alicudi (112–813 inhabitants in the same period). These are among the highest inbreeding coefficients observed in Italy.

# CHAPTER 8

# Effects of Inbreeding on Normal and Pathological Phenotypes

## 8.1 INTRODUCTION

The study of the progeny of consanguineous marriages can provide information on the phenotypic effects of inbreeding. Inbreeding will cause a relative increase of the frequency of homozygous individuals, and therefore is expected to reveal rare recessive traits. Sir Archibald Garrod used this principle to give the first proof of the validity of Mendel's rules for humans (1902). He showed that patients of certain recessive traits or diseases like albinism, alkaptonuria, and others, which he named "inborn errors of metabolism," have unusually high frequency in children of consanguineous parents. He correctly hypothesized that these defects were due to the lack of function of recessive genes controlling specific enzymes. Other than for this extraordinary example of intuition, humans were not the right organism for further developing the one gene–one enzyme theory. It was the work of George W. Beadle and Edward L. Tatum that, almost forty years later (1941), developed the theory that goes by this name, with experiments on the synthesis of vitamins and aminoacids in *Neurospora crassa*. S. Wright (1922) showed that the standard rule of distribution of the relative frequencies of homozygotes and heterozygotes expected under random mating, which goes under the name of Hardy-Weinberg, is modified under inbreeding: the frequencies of heterozygotes are decreased by the factor $1 - F$, where $F$ is the average inbreeding coefficient of the population, and those of homozygotes are increased accordingly (see chapter 10).

There are several ways in which research on inbreeding effects can be carried out. A retrospective one, which is most usually practiced, is to ask individuals, whether normal controls or patients, about possible consanguinity of their parents. This information is usually provided but there are two main reasons why it may be withheld: one is that many children of consanguineous parents may not know it or may not know it precisely. The other is that in many countries or social layers there are social objections to consanguinity, and individuals or patients may not be willing to acknowledge their inbreeding or their parents' consanguinity. There is thus a definite risk that consanguinity will be underreported.

In the investigations summarized in this chapter, we have almost always used lists of consanguineous marriages from Roman Catholic sources of the area under study, and linked them with the records of patients or normal subjects, of whom name and surname of father and mother were known. In the later surveys we had prepared an alphabetized list of about 540,000 consanguineous marriages that took place in Italy from 1914 to 1964. Even this may not be complete, and incompleteness may be feared especially when using partial lists of consanguineous marriages. For instance, in studies reported in this section, subjects were from the Province of Parma: the full list of Italian marriages was not yet available, but we could use that of the Diocese of Parma and neighboring ones for a perfectly adequate period of time. The probability that marriage occurred outside the neighboring dioceses and was consanguineous is to be considered, but must be rather low.

## 8.2 NORMAL QUANTITATIVE PHENOTYPES: STATURE AND CHEST GIRTH

The source of phenotypic data was measurements collected, for purposes of conscription, by the Italian Army, district of Parma, on all individuals born in the province in the years 1890 and 1891. Individuals were subjected to a highly standardized visit for ascertaining their fitness for military service, in the course of which a few measurements and other observations were taken. In particular, we used the measures of stature and that of chest girth, as well as other information available on the military records, such as place and date of birth, profession, and literacy.

In another investigation we could show the usefulness of record-linking these past records to other sources of information, such as the city population records, to find whether the same conscripts had married, emigrated, or died. We could show that there was no difference in mean age at death, probability of emigration or marriage for stature or chest girth, but both measurements showed a significant variation of the chance of marriage: individuals with extremely high or low statures or chest girth had a probability of marriages decreasing with difference from the mean, a classical example of stabilizing natural selection (Conterio and Cavalli-Sforza 1960). This result was confirmed by a similar research on Harvard students (Damon 1969).

Correlations of stature and chest girth with environmental variables showed values mostly significant and always positive of both traits with exact age (clearly, there is still some growth after conscription age, 18 years), year of birth (indicating a secular trend of increase in stature, which is very manifest in Italy and other European countries), altitude of the birthplace, and literacy. Stature was lower on average in the farming profession.

Linking these data with consanguinity showed a significant negative linear correlation but only for chest girth (figure 8.1, from Barrai et al. [1964]). The linear regression of stature on the inbreeding coefficient was negative but not significant, with a very strong deviation from linearity, caused by an unexpectedly high stature for children of very low degrees of inbreeding (third and 2½ cousins). The deviation was reduced but not eliminated after correction by linear regression with environmental variables. After correction the regression was slightly improved. The data also indicated correlations of order $+0.548$ between sibs and of $+0.875$ among twins (conscripts born on the same day, only about half of which are expected to be identical).

An analysis of variance showed that there is significant variation of stature or chest girth with sibship size. Moreover, the variance between sibships increased considerably with sibship size. There must be a strong interaction between these phenotypic traits (socioeconomic conditions and sibship size), which may be responsible for

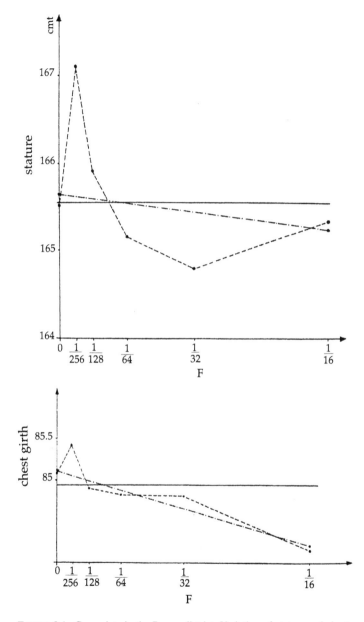

FIGURE 8.1. Conscripts in the Parma district. Variation of stature and chest girth as a function of inbreeding coefficient $F$.

the difficulties encountered with the high average phenotypic values observed with low inbreeding.

It seems reasonable to conclude that especially for chest girth and perhaps also for stature there is dominance (alleles causing high values of chest girth and probably also of stature are dominant), and therefore there is some inbreeding depression for these traits.

## 8. 3 MORTALITY, FERTILITY, AND STERILITY

The analysis of the effects of inbreeding on prereproductive mortality has shown the approximate validity of the expected relationship between them:

$$-\ln S_i = A + BF_i \qquad (8.1)$$

where $S_i$ is survival of individuals with $F_i$ inbreeding, and $A$ and $B$ are constants to which specific meanings can be assigned on simple hypotheses (Morton et al. 1956). Unfortunately, samples from single countries are usually of small size, and data from different countries or regions of a country cannot be safely cumulated for analysis because of heterogeneity of the socioeconomic factors affecting mortality levels.

Bittles and Neel (1994) have given a cumulative analysis of the correlation between percentage of mortality until 10 years of age (including very late abortions or stillbirths) in nonconsanguineous (non-C) offspring and that in the progeny of first cousins (1C), which is the most informative type of consanguineous marriages because of their frequency, availability, and genetic effect. The data come from published studies of 38 populations from around the world. The relationship is approximately linear and the calculation of the death rate due to first-cousin level of inbreeding ($F = 0.0625$) is easily obtained by extrapolating the death rate among the progeny of first cousins expected for a death of 0% in the non-C progeny. The result, 4.4%, is therefore an estimate of mortality in the progeny of first cousins due to inbreeding.

One may expect on the basis of (8.1) that the highest degree of inbreeding observed in human populations (uncle–niece or aunt–nephew or double first cousins) may cause a mortality slightly less than twice as large, while the mortality in the second most frequent type of consanguineous marriage (second cousins) is expected to be about 1%. These values may seem high in modern Western populations, in which the comparable death rate of nonconsanguineous marriages is of the order of 2–4%, varying somewhat with country. Thus, children of first-cousin marriages should have a death rate at least twice as high as that of noninbred individuals. This is based on present probabilities of death in Western countries, while over a century ago death rates were also much higher in Europe. Today, the majority of developing countries have spontaneous general death rates that vary from 10 to 40%, so that the mortality added because of parental consanguinity is hardly perceptible. This explains why there is no real fear of consanguineous marriage in most developing countries, but in developed ones the cost of inbreeding can be clearly perceived, at least for marriages of first cousins and of closer relatives.

Studies of fertility have given inconclusive results, as discussed by Bittles (1994). There are reasons why one might expect an increased fertility in inbred progeny because they are more similar immunologically. But one might also expect inbreeeding depression for genes controlling fertility. Available data do not seem to settle the issue.

Studies by Conterio and Barrai (1966) and Conterio (1967) have contributed data on the effects of inbreeding on mortality, sterility and fertility of Italian populations. One study was conducted in the Parma Diocese by taking 492 consanguineous families and 420 controls matched for age and social conditions. A study of mortality, sterility, and fertility is given in table 8.1. There is a fairly regular decrease of average number of pregnancies with increasing $F$, and the number of children born and surviving at reproductive age show the same pattern. These observations are significant, while sterility data show no clear effect, except for abortions plus stillbirth. These analyses depend very much on samples size, but the size of the Parma sample is close to the average of the samples selected by Bittles and Neel for their analysis of mortality data. When the Parma sample was compared with those listed by Bittles and Neel, total prereproductive mortality (stillbirths plus deaths until 20 years of age) in Parma was found to be 14.0% in the progeny of first cousins and 10.5% in non-C families. The difference, 3.5%, although not significantly different from zero, compares well with the estimate of

TABLE 8.1. Mortality, fertility, and sterility by inbreeding coefficient in the Parma Diocese

| | $F$ | | | | | | |
|---|---|---|---|---|---|---|---|
| | 1/8 | 1/16 | 1/32 | 1/64 | 0 | *Total* | *Significance* |
| No. of families | 6 | 127 | 80 | 279 | 420 | 912 | |
| No. of sterile families | 1 | 16 | 7 | 28 | 44 | 96 | |
| No. of pregnancies | 12 | 273 | 198 | 654 | 1093 | 2230 | |
| No. of live born | 11 | 230 | 164 | 582 | 966 | 1953 | |
| No. of abortions | 1 | 38 | 27 | 58 | 109 | 233 | |
| No. of dead before birth (abortions + stillbirths) | 1 | 43 | 34 | 72 | 127 | 277 | |
| No. of dead before 20 years age | 1 | 28 | 14 | 57 | 95 | 195 | |
| Rate of sterile families | 16.67 | 12.60 | 8.75 | 10.03 | 10.48 | 10.53 | <0.80 |
| Rate of abortions | 8.33 | 13.92 | 13.64 | 8.87 | 9.97 | 10.45 | <0.10 |
| Stillbirth rate | 8.33 | 15.75 | 17.17 | 11.01 | 11.62 | 12.42 | <0.10 |
| Postnatal mortality | 8.33 | 10.26 | 7.07 | 8.71 | 8.69 | 8.74 | <0.80 |
| Average no. of pregnancies | 2.00 | 2.15 | 2.47 | 2.34 | 2.60 | 2.44 | <0.02 |
| Average no. of liveborn children | 1.83 | 1.81 | 2.05 | 2.09 | 2.30 | 2.14 | <0.01 |
| Average no. of children in reproductive age | 1.50 | 1.61 | 1.94 | 1.86 | 2.06 | 1.92 | <0.01 |

*Source.* Conterio (1967).

4.4% given by Bittles and Neel of the mortality of progeny of first cousins.


## 8.4 INCIDENCE OF DISEASE GROUPS FROM SURVEYS OF HOSPITAL POPULATIONS

A survey conducted in the Nuoro University hospital archives, Sardinia, by Conterio (1967) record-linked the 9,636 hospital entries in years 1952–1957 with the list of consanguineous marriages of the nearby dioceses, from which patients are hospitalized at Nuoro. Sardinia has high consanguinity rates, and 1,234 (12.8%) of patients turned out to be inbred. This is more than three times the Italian average, and seven to eight times that of France and Belgium for comparable periods. The distribution of consanguinity degrees among patients was very close to that of the general population (table 8.2).

Diagnoses of patients' diseases that represented the main cause for hospitalization were sorted into nine classes according to the standard classification of the Italian Istituto Centrale di Statistica. Table 8.3 shows the average inbreeding coefficient of patients of the various disease classes. Highest inbreeding values are observed for metabolic, lymphatic and lymphopoietic system, and central nervous system diseases. Only the last are significant. Significance depends on numbers of observations, but this is not the only factor affecting the chance of reaching significance. Inbreeding tends to be especially high for very rare recessive diseases. If there are many different recessive diseases, each rare, in a population, the effect of consanguinity is going to be very high. But a frequent recessive disease, like thalassemia in Sardinia, will not show easily an increase of consanguineous parentage, as simple algebraic analysis can prove (see, e.g., Cavalli-Sforza and Bodmer 1971, 1999).

A similar survey was carried out in the University hospital of Parma by Conterio and Barrai (1966). Table 8.4 summarizes results for six defects selected among those that gave the greatest response to inbreeding or were especially interesting for other reasons. The number of "detrimental equivalents," that is, the number of detrimental genes per gamete, can be taken as approximately equal to the $B$ values (see equation 8.1) given with their standard errors in the table.

TABLE 8.2. Distribution of consanguinity degrees among 9,636 patients of the Nuoro hospital in the years 1952–1957 (numbers and percentages), compared with percentages of consanguineous marriages in the Nuoro population, 1875–1957

| | F | | | | | | | Multiple consanguinity | Nonconsanguineous |
|---|---|---|---|---|---|---|---|---|---|
| | 1/8 | 1/16 | 1/32 | 1/64 | 1/128 | 1/256 | | | |
| No. of hospital entries | 2 | 271 | 135 | 414 | 125 | 185 | | 102 | 8,402 |
| % consanguineous | 0.16 | 21.96 | 10.94 | 33.55 | 10.13 | 14.99 | | 8.27 | |
| % consanguineous in Nuoro population | 0.19 | 25.00 | 10.36 | 31.52 | 9.58 | 13.96 | | 9.39 | |

*Source.* Conterio (1967).

TABLE 8.3. Average inbreeding coefficient $(F)$ of patients of the Nuoro hospital in 1952–1957 for the nine groups of diseases compared for significance with the average inbreeding coefficient of the general population $(\overline{F})$ by a $t$ test

| Disease | $F \times 10^{-3}$ | $\sigma^2 F \times 10^{-3}$ | $\overline{F} \times 10^{-3}$ | $F/\overline{F}$ |
|---|---|---|---|---|
| Pulmonary tuberculosis | 3.42 | 0.1302 | 4.36 | 0.78 |
| Viral diseases | 5.30 | 0.2169 | 4.20 | 1.26 |
| Tumors of lymphatic and lymphopoietic system | 8.07 | 0.6295 | 4.51 | 1.79 |
| Endocrine diseases and diabetes | 5.45 | 0.2154 | 4.57 | 1.19 |
| Metabolic diseases | 8.27 | 0.4307 | 4.21 | 1.96 |
| Hematological diseases | 5.59 | 0.2534 | 3.89 | 1.44 |
| Thalassemia | 4.94 | 0.1475 | 4.01 | 1.23 |
| Nervous central system diseases | 7.06 | 0.2762 | 4.13 | 1.71** |
| Congenital malformations | 5.83 | 0.2527 | 4.20 | 1.3 |

Source. Conterio (1967).
**Significantly different from zero at $P = 0.01$.

The significance of the linear regression and that of deviations from linear regression are also shown. Tuberculosis and malformation, not significant in this survey, were also not significant in the Nuoro survey. All the other four classes of disease show a significant linear regression with the degree of consanguinity and, thus, the existence of recessive detrimental genes. There is no significant deviation from linearity except for one of the four classes, mental retardation. This is probably due to the relatively high frequency of nongenetic and non-recessive causes of mental retardation.

A survey of consanguinity in mental disease was carried out by Conterio and Zei (1964) in the psychiatric hospitals of Parma, Piacenza, and Reggio Emilia (the dioceses of northern Emilia most thoroughly investigated for consanguineous marriages) between 1947–48 and 1960–61. The frequency of patients who originated from consanguineous parents was 2.68% in Colorno (Parma), 2.71% in Reggio Emilia, and 2.86% in Piacenza. As discussed in chapter 3,

TABLE 8.4. Detrimental load and equivalents by type of defect, from a survey in the Parma hospital

| Type of defect | A | B | B/A | Regression on F | Deviation from linearity |
|---|---|---|---|---|---|
| Mental retardation | 0.024 | 1.399 ± 0.3 | 58.3 | *** | ** |
| Eye defect | 0.058 | 0.823 ± 0.3 | 14.1 | * | — |
| Ear defect | 0.013 | 0.565 ± 0.2 | 41.8 | ** | — |
| Tuberculosis | 0.020 | 0.198 ± 0.2 | 9.6 | — | — |
| Malformations | 0.040 | 0.487 ± 0.3 | 12.1 | — | — |
| Smaller defects | 0.081 | 1.082 ± 0.4 | 13.3 | ** | — |

Source. Conterio and Barrai (1966).
*Significantly different from zero at $P = 0.05$.
**Significantly different from zero at $P = 0.01$.
***Significantly different from zero at $P = 0.001$.

the three provinces are particularly homogeneous ecologically and from a socioeconomic point of view.

Table 8.5 shows the distribution of the numbers of patients for the six major mental diseases by inbreeding coefficient, the average degree of inbreeding of patients, and the comparable value for the general population, taking into account the age of patients and therefore the presumptive distribution of age at marriage of parents. In this area, however, the frequency of consanguineous marriages has been relatively stable in time and started decreasing seriously only after 1920. The average inbreeding of patients is clearly greater than that of the general population only in the case of oligophrenia, which is frequently the consequence of many rare, inherited, recessive genes. Penrose (1963), however, noted that at least in 15% of the cases the disease is caused by prenatal or birth trauma, or encephalitis. It is possible that for young patients the average inbreeding calculated for the general population is overestimated, and hence the increase of inbreeding among patients is underestimated.

In all other diseases there is less effect of consanguinity than expected at random. This could be observed if genetic dominant diseases are involved, but the differences are not significant. In schizophrenia the inbreeding coefficient of patients and of the general

TABLE 8.5. Distribution of numbers of patients of northern Emilia psychiatric hospitals (1947–61) by type of mental disease and by inbreeding coefficient

| Mental disease | $\frac{1}{16}$ | $\frac{1}{32}$ | $\frac{5}{64}$ | $\frac{1}{64}$ | $\frac{1}{12}$ | $\frac{17}{256}$ | $\frac{9}{256}$ | $\frac{3}{256}$ | $\frac{1}{256}$ | 0 | Totals | $F \times 10^{-3}$ | $\bar{F} \times 10^{-3}$ | $F/\bar{F}$ |
|---|---|---|---|---|---|---|---|---|---|---|---|---|---|---|
| Schizophrenia | 12 | 4 | 1 | 11 | 1 | 1 | — | 2 | 7 | 985 | 1,024 | 1.22 | 1.19 | 1.025 |
| Manic depressive psychosis | 4 | 1 | — | 10 | 7 | — | 1 | — | 5 | 1,049 | 1,077 | 0.51 | 1.36 | 0.375 |
| Alcoholism, alcoholic psychosis | 1 | 2 | — | 3 | 2 | — | — | — | 5 | 493 | 506 | 0.41 | 1.29 | 0.318 |
| Senile psychosis | 1 | 3 | — | 4 | 1 | — | — | — | 1 | 195 | 205 | 1.12 | 1.30 | 0.862 |
| Psychoneuroses | 4 | 5 | — | 13 | 4 | — | — | — | 8 | 1,378 | 1,412 | 0.48 | 1.27 | 0.378 |
| Oligophrenia | 17 | 3 | — | 6 | — | — | — | — | 4 | 535 | 565 | 2.24 | 1.23 | 1.821* |
| Total | 39 | 18 | 1 | 47 | 15 | 1 | 1 | 2 | 30 | 4,635 | 4,789 | | | |

*Source.* Conterio and Zei (1964).

*Note.* Average inbreeding coefficient of the patients ($F$) is compared with the average inbreeding coefficient of the general population ($\bar{F}$) by a *t*-test.
*Significantly different from zero at $P = 0.05$.

population is the same, and the presence of recessive genes is not confirmed.


## 8.5 STUDY OF SPECIFIC RECESSIVE DISEASES

A further extension of the theory of recessive mutations involves the study of specific recessive diseases. Dahlberg (1948) gave a simple theory proving what the intuition of Garrod had foreseen, namely that the frequency of recessive defects is increased in the children of consanguineous matings. His formula gives the frequency $C'$ of first-cousin marriages among the parents of homozygotes for a recessive trait, if the frequency $C$ of first-cousins in the general population and the frequency $q$ of the recessive gene in the general population are known:

$$C' = \frac{C(1+15q)}{16q + C(1-q)} \qquad (8.2)$$

It can easily be shown that $C'$ is always greater than $C$, the more so, the smaller is $q$ (see figure 8.2). From this formula one can use the observed frequencies of $C$, $C'$ to calculate the frequency of a recessive gene:

$$q = \frac{C(1-C')}{16(C'-C) + C(1-C')} \qquad (8.3)$$

Because the relative frequency of homozygous recessives, in the absence of natural selection, is $q^2$, it is very small for defective traits, which are kept into low frequencies by natural selection. At equilibrium between mutation $\mu$ toward a deleterious recessive gene, and the intensity $s$ of natural selection against it, the expected frequency of the homozygous genotype (its "incidence"—also called "prevalence," two terms that have different meanings and are sometime used improperly) is equal to $\mu/s$. If selection against the homozygous genotype is complete, $s = 1$, that is, for recessive lethals, and the expected frequency of the recessive trait is equal to the mutation rate $\mu$, explaining why it is very low. This estimate implies the absence of

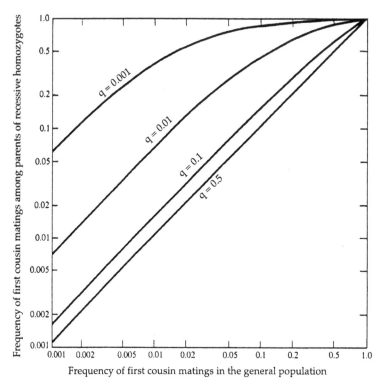

FIGURE 8.2. Frequency of first-cousin marriages expected among the parents of progeny displaying recessive phenotypes. *From Cavalli-Sforza and Bodmer [1971, 1999].*

heterozygous advantage. If selection happens after birth, and the relative frequency of homozygous recessives at birth is known (that is, their frequency before the action of selection), one can calculate the frequency of the recessive allele by taking the square root of the frequency of the defect at birth. Since the defective individuals are very few, however, the estimate of $q$ is affected by a large error (Cavalli-Sforza and Bodmer 1971, 1999, pages 88–92).

The incidence of very rare diseases is difficult to estimate in the absence of general population surveys. There is always a chance that some cases will be lost, and, unless the responsible gene is well known and has been tested in all the families of the cases that have been observed, there may be heterogeneity of the disease. The method of estimating the gene frequency of recessives from consanguineous matings is useful when $q$ is low, that is, when the defect is rare. The simplest procedure is to estimate the frequency of first cousins among the affected and that in the general population. The gene frequency $q$ of the recessive gene is then given by formula 8.3, and its square is the expected incidence of the disease. The estimation using only first cousins is simple and covers most of the information available, but it is possible to improve the precision of the estimate of $q$ by using all consanguineous matings, with more general formulas (see Chung et al. 1959, Barrai et al. 1965). Standard errors of estimates of $q$ and of the incidence of the trait can then also be computed (Barrai et al. 1965). The formulas are

$$q = \frac{V}{F - \alpha} - \alpha; \qquad SE(q) = \left[ \frac{V}{(F - \alpha)^2} \right] SE(F) \qquad (8.4)$$

where $\alpha$ is the average inbreeding coefficient of the population, $V$ its variance, $F$ the average inbreeding coefficient of the affected individuals, and SE refers to the standard error of the quantity in parentheses after it.

We describe the results of seven investigations on recessive diseases in Italy, in which estimates were obtained by making resort to the alphabetical list of consanguineous mates established in Parma from the Vatican archives. The consanguinity frequencies of the general population were thus calculated from these archives and extrapolated by a negative exponential according to the pattern of change in time for years after 1964, the last year for which the data are available in the archives. Information on parental consanguinity was obtained directly from the patients' families.

The diseases, the number of families $N$, the geographic coverage of the investigations, and the relevant publications and dates are as follows:

- Cystic fibrosis: $N = 624$, all Italy, Romeo et al. (1985)
- Friedreich ataxia: $N = 83$, all Italy, Romeo et al. (1983a)
- Friedreich ataxia: $N = 59$, Northwestern Italy, Leone et al. (1990)

- Phenylketonuria (PKU): $N = 178$, all Italy, Romeo et al. (1983b)
- Ataxia telangectasia: $N = 72$, all Italy, Chessa et al. (1994)
- Werner syndrome: $N = 3$, Sardinia, Cerimele et al. (1982)
- Hemophagocytic lymphohistiocytosis: $N = 67$, all Italy, Aricò et al. (2000).

Table 8.6 gives the values of $C'$, $C$ of first cousins, $\alpha'$ ($= F$), and $\alpha$ of average inbreeding for the parents of patients and for marriages of the general population, respectively, and the estimate of $q$ derived from the two methods.

Application of this method allows an estimate of the general incidence, which, especially for rare diseases, is extremely difficult otherwise. When there is some reasonable estimate of the incidence from direct observations, which happens mostly only for less rare diseases, one can compare it with the expected one, thus obtaining information on the possible existence of more than one gene determining the disease. If the genes are widely separated on the same chromosome or in different chromosomes, this information could also come from linkage studies. In the six diseases listed above there is most probably a single locus, since the evaluation of the incidence is in reasonable agreement with that expected for a single locus.

In the case of cystic fibrosis doubt had been expressed on the hypothesis that a single locus is involved, because the pathological picture is variable and there are two major types of pathologies, respiratory and intestinal. Others (see Romeo et al., 1983a) have suggested the existence of more than one gene for other reasons. In general, however, the incidence of the diseases, 1/2,400, estimated with the screening methods then available, was in very reasonable agreement with that estimated by the Barrai method ($\alpha'/\alpha$), which is 1/2,367. The Dahlberg method, based on first-cousin frequencies only, gave a lower value, 1/1,635. The frequency of the disease was thus estimated to be around 1/2,000. This is incompatible with the hypothesis that there exist two genes of equal frequency. At the time, there existed a neonatal screening in the Veneto region based on 229,626 newborns, tested in 8 years. The information on parental consanguinity of the probands and their affected sibs for the 624 cystic

TABLE 8.6. Data bearing on the relationship of consanguinity and six
recessive diseases

| Disease | $C'\%$ | $C\%$ | $\alpha'$ | $\alpha$ | $q(C)$ | $q(\alpha')$ |
|---|---|---|---|---|---|---|
| Cystic fibrosis | 2.40 | 0.71 | 0.00190 | 0.00054 | 0.02473 | 0.02055 |
| Friedreich ataxia | 14.50 | 1.62 | 0.00979 | 0.00118 | 0.00671 | 0.00632 |
| Friedreich ataxia | 3.33 | 0.67 | — | — | 0.00523 | — |
| Phenylketo-nuria | 7.87 | 0.96 | 0.00512 | 0.00057 | 0.00801 | 0.00749 |
| Ataxia tel-angectasia | 6.94 | 1.23 | 0.00586 | 0.00099 | 0.01234 | 0.00963 |
| Werner syndrome | 42.86 | 3.62 | 0.03571 | 0.00316 | 0.00329 | 0.00148 |
| Hemophagocy-tic lympho-histiocytosis | 14.92 | 1.04 | — | — | 0.00396 | — |

Note. $C'$ = frequency of first-cousin matings among the parents of the patients, $C$ = frequency of first-cousin matings in the general population, $\alpha'$ = average inbreeding among the parents of the patients, $\alpha$ = average inbreeding in the general population, $q$ = estimated gene frequency: $q(C)$ from first cousins, $q(\alpha)$ from all consanguinity degrees.

fibrosis (CF) sibships came from 11 CF centers of Verona, Milan, Genoa, Parma, Turin, Trieste, Cesena, Florence, Rome, Naples, and Bari, representing most of Italy except the major islands.

The state of CF is much better known today because the responsible gene, CFTR, has been identified. The number of known mutations causing CF is over 800, but the gene remains unique. The frequency of the various pathogenic alleles varies geographically, the major one (DF508) being responsible of about half of all Italian cases. Its relative contribution also varies geographically, being highest at the periphery of the Neolithic expansion and smaller in southeastern Europe. In spite of all the DNA knowledge, the best estimate of the incidence of CF remains 1/2,000, the estimate obtained by Romeo et al. in the consanguinity study published in 1985.

For Friedreich ataxia doubts had also been expressed that only one gene was involved. Of the 83 families of the study of Romeo et al. (1983a), 61 came from northern Italy, 5 from central Italy, and 17 from southern Italy. A number of patients living in northern Italy are of recent southern origin, as a consequence of permanent or temporary immigration. It is possible that the disease is more frequent in southern Italy, although the data are not sufficient to decide. Since southern Italy was largely settled from the southern Balkans in Neolithic times, and again from Greece in the first millennium B.C., it would be interesting to test the disease incidence in the areas of origin of southern Italians and compare it with that of northern Italians. There is an estimate of prevalence of the disease in the Campania region of 1/90,901. Taking account of the expectation of life of Friedreich patients (30 years) versus that of the general population (70 years), one can calculate approximately the incidence at birth to be 70/30 × prevalence, which is equal to 1/38,958. This is not seriously different from the incidence in southern Italy from all consanguinity values, 1/28,392. Thus, here also there is no reason to doubt that there is a single gene of Friedreich ataxia. The results obtained by Leone et al. (1990), based only on first-cousin analysis, led to the same conclusion.

Phenylketonuria (PKU) is systematically screened on newborns because dietetic treatment, started at birth, can avoid permanent brain injury. Unfortunately, screening is not always practiced generally, especially in the south, but at least it is sufficiently frequent that it generated good estimates of incidence, separate from those of hyperphenylalaninemias, which is almost twice as frequent. There is very little difference in the PKU frequency of Italian regions, with mean incidence of 1/12,000. The only important variation is in the Venetian region and Friuli (Udine), which is almost three times lower (1/32,439).

Ataxia telangectasia shows very little if any difference in the estimated incidence between northern and southern Italy, with an average of 1:7,090. It is possible that the incidence is higher, since it became twice as high in later years, when the diagnosis became easier.

Werner syndrome is very rare. The 10 cases used for this analysis all come from 3 families of a central region of Sardinia, the province of Nuoro. The estimates from first cousins and all consanguinities vary from 1/92,515 to 1/454,505. Perhaps the right frequency is like that estimated in Japan, 1/300,000.

Hemophagocytic lymphohistiocytosis (HLH) shows a great geographic heterogeneity of the 92 cases ascertained in the period 1976–

1999, its frequency being over three times rarer in the northern Italy than in the south, where 50 of the 67 HLH families are present. Moreover, the high concentration of HLH in the two "Magna Grecia" regions of Campania and Sicilia (43/50) suggest a possible heterogeneity of the disease, to be studied.

# CHAPTER 9

# Consanguineous Marriages in Italy: Data from the Vatican Archives

## 9.1 INTRODUCTION

"Consanguineous marriages in Italy: information gathered from the matrimonial dispensations requested in 1911–1964 of the Sacred Congregation of the Sacraments, stored in the Vatican's Secret Archives." This is the title of a set of 44 volumes produced by Antonio Moroni's work in the 1960s, from which one can study the frequencies, causes, and effects of consanguinity in the whole of Italy over a 54-year period.

The following information was magnetically filed from each of 520,492 consanguineous marriages for which dispensation was requested:

- Surname and Christian name of husband and wife
- Age at marriage
- Year of marriage
- Name of diocese where the marriage took place
- Degree of consanguinity of the couple
- Type of genealogical tree of consanguinity (not always available)
- Dispensation granted—free or against payment.

Permission was obtained from the Catholic authorities to record-link this file with information containing paternity and maternity and other larger data, especially of normal and pathological phenotypes of various sources. This chapter analyzes time and space variation of consanguinity and its correlation with ecological and socioeconomic factors obtained from the Italian Istituto Centrale di Statistica for comparable periods and from other sources.

## 9.2 VARIATIONS OF CONSANGUINITY OVER TIME

A general view of consanguinity trends over the whole period under consideration may be obtained from the average inbreeding coefficient $\alpha$ calculated as an average of the frequency of different degrees of consanguineous marriages, $p_i$, weighted by the inbreeding coefficient $F_i$ (see figure 1.1):

$$\alpha = Sp_iF_i \tag{9.1}$$

Plotted year by year, as in figure 9.1, $\alpha$ generally reduces with time, starting from the highest value of $2.48 \times 10^{-3}$ in 1919, down to $0.76 \times 10^{-3}$ in 1961. The major irregularities in an otherwise uniform pattern were observed during the World Wars of 1914–1918 and 1940–1945. During these periods consanguineous marriages became less frequent, compared with the total numbers of marriages. In the immediate postwar periods they increased in number to figures higher than before the war. Perhaps marriages were postponed in wartime, due to the difficulty in obtaining dispensations for marriages between relatives; others got married without declaring they were related. Yet other less easily checked reasons might be suspected for this phenomenon, for example, when men returned from the war the reestablishment of family contacts could have encouraged intrafamily marriage. After all, a major reason to marry within the family is that the reciprocal knowledge of a member of the family is greater than that of more superficial, recent acquaintances, and a soldier returning from war may desire to establish a marriage relationship soon. A very familiar person may be a more desirable mate for both man and woman.

We shall deal with four major types of consanguineous marriages: uncles–nieces and aunts–nephews, first cousins, first cousins once removed, and second cousins (listed in decreasing order of $F$). For more distant consanguineous relationships, matrimonial dispensation has not been necessary since 1917.

Considered separately, the frequency of the four major types of

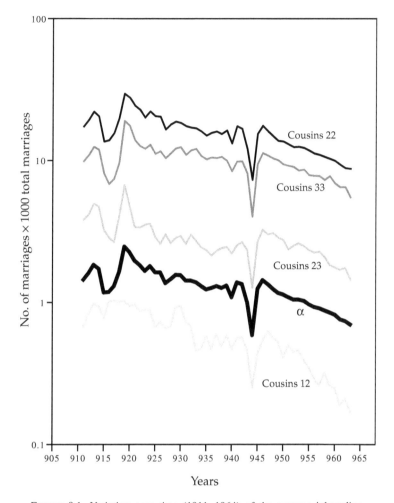

FIGURE 9.1. Variation over time (1911–1964) of the average inbreeding coefficient α and the four types of consanguineous marriages: uncle–niece/aunt–nephew (12), first cousins (22), first cousins once removed (23), and second cousins (33).

cousins per 1000 marriages (see figure 9.1) reveals a very similar trend, which obviously ties up with α. The various levels show clear frequency differences, in agreement with the observations from data collected in the bishoprics and studied in earlier chapters: even degrees (first and second cousins) are most frequent; marriages with

TABLE 9.1. Contribution of different types of consanguineous marriage to
the value of $\alpha$: Italy 1911–1961, averages

| Type of marriage | F | $p_i F_i$ | % |
|---|---|---|---|
| Uncle–niece, aunt–nephew | 1/8 | 0.0746 | 5.54 |
| First cousins | 1/16 | 1.0203 | 75.70 |
| First cousins once removed | 1/32 | 0.0904 | 6.71 |
| Second cousins | 1/64 | 0.1625 | 12.05 |
| $\alpha$ | | 1.3478 | |

uneven relatives (first cousins once removed, uncles–nieces and aunts–nephews) are less frequent, in this order.

If the relative contribution given by each type of marriage to the value of $\alpha$ is measured, we can observe (table 9.1) that the total of first and second cousins accounts for approximately 88% of the value of $\alpha$; first cousins alone make up over 75%, and can describe with a reasonable approximation not only the trend but also the degree of consanguinity in Italy in the first half of the century.

## 9.3 GEOGRAPHICAL VARIATIONS:
## PROVINCES AND REGIONS

The only geographical indication on the dispensation kept in the Vatican archives as to where consanguineous marriages took place is the *diocese*. There are 280 dioceses in Italy at the time of this analysis. They constitute an ecclesiastical territorial subdivision, which bears no relation to the administrative ones: 92 *provinces*, subdivided into about 8,000 *communes* ("comuni"), each of which makes annual returns of civilian status (births, marriages, and deaths). The number of communes varies somewhat from year to year, as a few new ones are created and old ones suppressed. Each diocese is made up of parishes, which, from the administrative point of view, belong to a certain number of communes. Almost always, one commune consists of parishes from the same diocese, but a few communes are made up of parishes belonging to different dioceses.

This lack of correspondence between ecclesiastical and administrative boundaries made adjustments necessary to reduce the former to the latter. We used the data from the 1951 population census, provided by the Istituto Centrale di Statistica, in which parishes are listed by diocese, by commune, and by demographic size. First we grouped the approximately 8,000 communes or subdivisions of communes according to dioceses. Then the frequency of consanguineous marriages within the 92 Italian provinces was divided proportionately, according to the number of inhabitants in the communes that made up the province. This introduced a small error, but was necessary to compare consanguinity frequencies with demographic and socioeconomic data available for the whole province and the period under consideration.

The above-mentioned reduction in geographical units—from 280 dioceses to 92 provinces—and the consequent division of consanguineous marriage frequency figures enabled us to analyze spatial consanguinity variation using more satisfactory territorial units. In fact dioceses vary in size considerably between the north and south of Italy for a variety of historical reasons, and frequently have boundaries dictated by old and presently less meaningful criteria. Moreover, only at the province level are the total marriage figures available against which consanguineous marriage figures can be compared. Figures 9.2–9.5 show the frequencies by province of the various types of consanguineous marriage at three different times: the beginning, middle, and end of the 50 years under review (data of second cousins for the island of Sicily were not available in the Vatican archives). There are great frequency differences of the various types of marriage in the different provinces within one time period.

The figures show that adjacent provinces often have the same frequency density. Analyzing the behavior of each province over the whole review period enables us to define limits of relatively homogeneous areas. Elements common to all provinces are the start of a declining trend in consanguinity between 1915 and 1920 and the shape of the curve, approximately a negative exponential.

## 9.4 SPACE–TIME ANALYSIS: FOUR MODELS OF DECLINING CONSANGUINITY IN ITALIAN REGIONS

Apart from a few exceptions, provinces belonging to the same administrative region show declining consanguinity coefficient patterns, which can nearly be superimposed.

Uncle-niece and Aunt-nephew

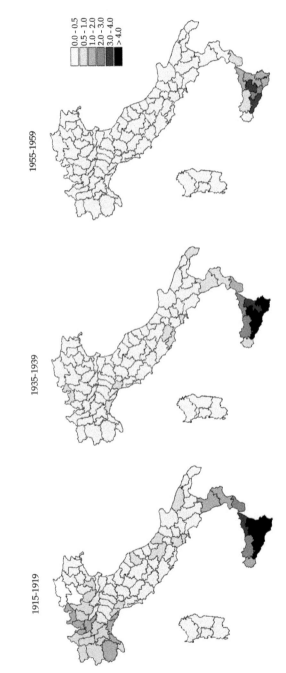

FIGURE 9.2. Variation of frequency of consanguineous marriages in Italian provinces at three different times: uncle–niece and aunt–nephew.

First cousins

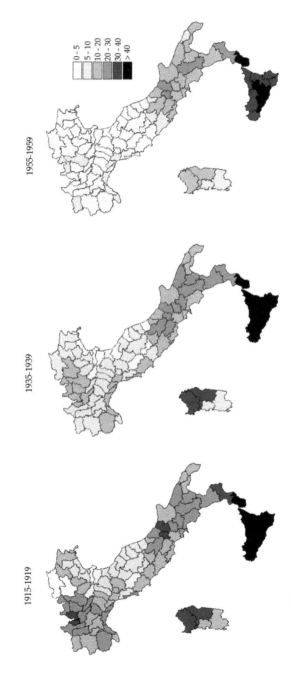

FIGURE 9.3. Variation of frequency of consanguineous marriages in Italian provinces at three different times: first cousins.

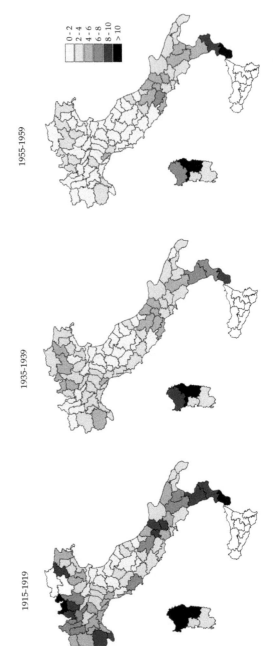

FIGURE 9.4. Variation of frequency of consanguineous marriages in Italian provinces at three different times: first cousins once removed.

Second cousins

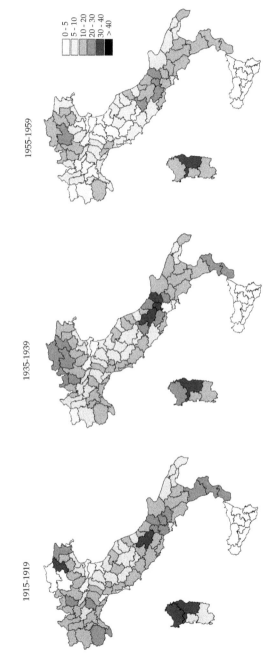

1915-1919        1935-1939        1955-1959

0 - 5
5 - 10
10 - 20
20 - 30
30 - 40
> 40

FIGURE 9.5.  Variation of frequency of consanguineous marriages in Italian provinces at three different times: second cousins.

Moreover, by principal component analysis (PCA) (see Cavalli-Sforza et al. [1994], pp. 39–42) of the average consanguinity figures of the 18 Italian regions, in time periods of 5 years (or those of first-cousin marriages, which we have seen are responsible on average for 75% of the $\alpha$ values), we have been able to group adjacent regions into four areas of apparently very different consanguinity trends (figure 9.6):

1. Northwest area (Piedmont, Lombardy, and Liguria, called the "industrial triangle")
2. Northeast and central area (Venetia, Friuli, Emilia, Tuscany, Marche, Latium, Umbria, and Abruzzi)
3. South (Molise, Basilicata, Campania, Apulia, Calabria, and Sardinia)
4. Sicily

There is a general decrease with time of the frequency of consanguineous unions in all areas, but the speed of descent is very different (figure 9.7). The northwestern region and the southern region, which at the beginning both revealed high average inbreeding coefficient values, at the end of the period (1961–1964) held, respectively, the lowest and highest values within the range of variation. On the other hand, the northeast and central Italian area always presents low average inbreeding levels and a slow rate of decrease.

In overall Italian data (figure 9.1), the relation between first and second cousins seems constant over time—both descending curves are virtually parallel. Within various provinces or regions, however, they are different and the analysis of the similarity of the joint trends of first and second cousin over time allows us to group the provinces into four distinct types roughly corresponding to the zones described above (see figure 9.7):

1. Northwest, where first-cousin marriages, more frequent in the first decade of the century, were (at different times in different provinces) overtaken by second cousins
2. Northeast and central, where second-cousin marriage is more common over the whole period
3. Southern, where first-cousin marriages are always more frequent than second ones

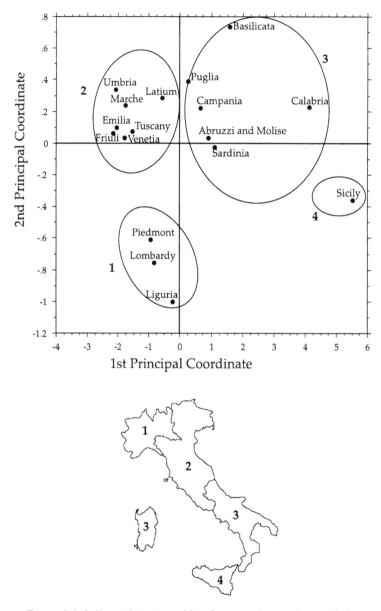

FIGURE 9.6. Italian regions grouped into four areas in accordance with the results of principal component analysis on time–space consanguinity values (α or first cousins).

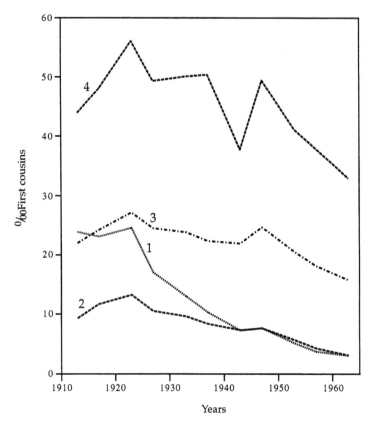

FIGURE 9.7. Trend of consanguinity decline in the four areas described by principal component analysis: 1, northwest; 2, northeast and center; 3, south and Sardinia; 4, Sicily.

4. Southern Calabria, which is similar to the third zone, except for a constantly higher ratio of first to second cousins.

## 9.5. FACTORS RESPONSIBLE FOR SPACE AND TIME DIFFERENCES IN CONSANGUINITY: CHOICE OF VARIABLES AND THEIR MEANING

Leaving out the effect of a population's particular attitude in favor of or against all or some types of marriages between relatives, the frequencies of consanguineous unions depend, primarily, on the "availability" of cousins to marry within the "genetic isolate," that is, on population size, fertility, mortality, and migration rates. After all, consanguinity depends on the *demographic structure* of an area that, in turn, is greatly affected by variation in the *socioeconomic structure*. Socioeconomic factors are responsible for influencing those social and cultural characteristics that caused the fertility rate to decline through restricting the number of children (birth control) and, hence, the reduced availability of cousins. Furthermore, economic factors regulate the migration rate and, hence, the dispersion of cousins and changes in isolate size. Therefore, a study of changes in population characteristics relating to the demographic, socioeconomic, and cultural situation of an area could illustrate the influence they have had on the frequency of consanguineous marriages and on the *genetic structure* of the Italian population in general.

From the whole series of statistics available per province, published by the Istituto Centrale di Statistica or deduced from the ten-yearly population censuses, we have selected the following:

- *Demographic* (birth, death, stillbirth, infant mortality, internal and external migration rates)
- *Economic* (number of people employed in agriculture and industry) and *social* (illiteracy)
- *Geographic* or *ecological* (altitude and population density)

From analyzing these aspects, which could be called *eco-social variables*, we can clarify causal factors of the considerable variations in consanguinity, since

- Different factors may have been more or less important in determining not only the mean value of the inbreeding coefficient in a region, but also the frequency of various types of marriages
- Changes in time of factors influencing the frequency of consanguineous marriages could have determined different patterns of consanguinity's variation in different areas.

## 9.6 DEMOGRAPHIC VARIABLES: BIRTHRATE, DEATH RATE, AND DEMOGRAPHIC TRANSITION

The relation between consanguinity and birth or death rates takes a different significance depending on whether these two purely demographic variables are measured at the time of consanguineous marriage or a generation or more earlier. In the first case, the birthrate particularly gives an indication of the degree of development of an area. This is generally associated with, or even determined by, socioeconomic factors.

In the second case, the birth/death rates of previous generations affect the frequency of consanguineous marriages through the relative "abundance" of cousins. In a growing population, the relative frequency of first cousins increases almost with the square of sibship size and, therefore, it increases with the rate of population increase, with a delay of one generation. Similarly, the higher the number of births in the grandparental generation, the greater the number of second cousins. So, in the final analysis, the progress of consanguinity over time could be, in part, determined by the increment rate of the population as a function of the difference between birth and death rates.

In Europe there was a complex decrease of population increment rates, beginning mostly in the nineteenth century, referred to as the "demographic transition." In connection with the "Princeton European Fertility Project," demographic transition, viewed as a decline in fertility, has been studied in the main European regions (*The Decline of Fertility in Europe*, Coale and Watkins, eds. [1986]). In Italy, the demographic transition was studied by Livi-Bacci (*A History of Italian Fertility*, [1977]); the transition was later than in the northern European countries, and differed greatly in the various regions.

Because of this, we deemed it necessary to study the demographic transition in Italy in the various provinces and regions and correlate it with the change in consanguinity frequencies in the period of interest to us. From information we have gathered in earlier chapters, summarized in section 9.8, we know that there was a substantial increase in Italian consanguinity in the first part of the nineteenth century that continued until about 1915–1920, after which there began a decrease in almost all areas. The Vatican data refer mostly only to the second period, in which there was a marked decrease. Nevertheless, we will try to consider in the last section of this chapter the general picture of consanguinity in the last two centuries. In this section we study the regional variation of the time trends of the demographic transition in Italy.

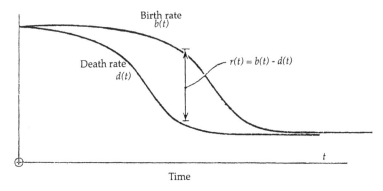

FIGURE 9.8. Theoretical model of demographic transition in Europe. *From Keyfitz [1985].*

Numbers of births and deaths per province have been registered in the Annuari di Statistiche Demografiche, since 1862. These data are collected every ten years directly from the population census or calculated annually from population movement figures. Hence, we have been able to estimate annual birth and death rates for every province and analyze the varying curves of demographic transition in Italy over a whole century (1862–1961).

In the theoretical European model (figure 9.8) birth and death rates start at about the same level, the death rate starts going down almost linearly with time, the birthrates start going down with a shorter or longer delay with about the same slope as the death rates, and both again become constant in time at the same level at the end, so that both before and after the transition the population is almost stationary in numbers. But in Italy, in 1862, the birthrate was higher than the death rate in all provinces (though it is not known whether, before that date, there was a period when they were equal). Especially in the subsequent decades, strong emigration became common to resolve economic problems due to the excessive growth of the population, especially in the less-developed parts of the country. Moreover, in the intervening 100 years, the provinces show considerable differences in the time at which the birthrate began to decline, in the slope of the regular death rate decline, and also in the absolute values of the rates.

Principal component analysis of the time–space values of birth

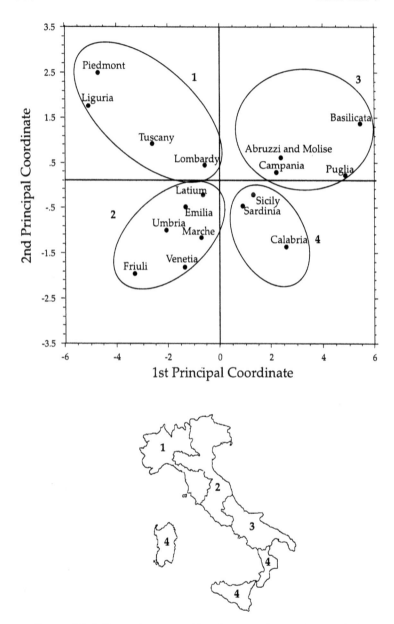

FIGURE 9.9. Italian regions grouped into four areas in accordance with the results of principal component analysis on time–space values of birth and death rates. 1, northwest; 2, northeast and center; 3, south; 4, Calabria, Sicily, and Sardinia.

and death rates showed considerable difference between provinces, but provinces belonging to the same region had very similar trends. Groups of adjacent geographical regions form four distinct areas (figure 9.9):

1. Northwest (Piedmont, Liguria, Lombardy, and Tuscany)
2. Northeast and central (Venetia, Friuli, Emilia, Latium, Umbria, and Marche)
3. South (Abruzzi and Molise, Basilicata, Apulia, and Campania)
4. Calabria, Sicily, and Sardinia.

A schematic representation of the pattern of demographic transition shared by the provinces in each of the four areas is seen in figure 9.10. Correspondingly, different trends of population rate of increase over time can be calculated for each of the four major models: from the time behavior of the difference between birthrate and deathrate, they were decreasing, parabolic, constant, and increasing, respectively.

Leaving aside all other factors, especially the socioeconomic ones, which caused consanguinity to decline in the first 50 years of this century, each of the four models of consanguinity trends previously studied (section 9.4) corresponds approximately to a specific demographic transition model.

The effect of the demographic transition on number of cousins depends on the variation of the increment rate of the population. If the increment rate of the population is constant in time, as is true in the Italian southern regions in most of the period, the relative proportion of cousins should be at an equilibrium value, and does not change from demographic considerations alone. In the extreme south there was some increase with time of the rate of population increment, but it was probably balanced by transatlantic emigration. Thus, in both cases we do not expect much change in consanguinity rate with time because of the demographic transition. In the central and northern regions the increment rate is decreasing, and thus one would expect a decrease of the proportion of relatives, and, therefore, of consanguinity, rather than an increase. Thus, the demographic transition may have caused a definite increase of consanguinity in the south, a lesser effect in the same direction in the center, and still less in the north.

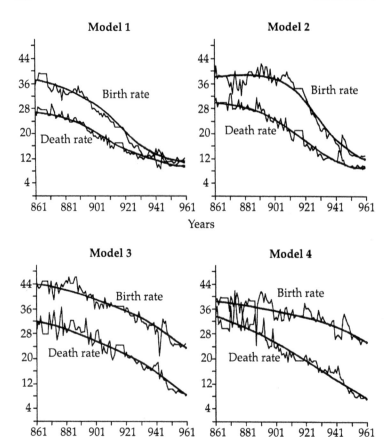

FIGURE 9.10. Schematic representation of the model of demographic transition shared by the provinces in each of the four areas described in figure 9.9.

However, the effect of the rate of population increment over time on the trend of consanguinity is subject to a delay of time of one generation for first cousins and two generations for second cousins. This, and the inevitable interaction with other socioeconomic changes, makes a precise analysis difficult, and we will be forced to limit ourselves to semiquantitative conclusions in the last section of this chapter.

## 9.7 EFFECT ON CONSANGUINITY OF ENVIRONMENTAL VARIABLES OF SOCIOECONOMIC AND ECOLOGICAL MEANING

In 1911 Italy was divided into 69 provinces. From 1927 to 1934 twenty-three new provinces were created by removing territory from one or more of the already existing provinces. Because a comparison between the two periods requires the same territorial subdivisions, the 92 provinces existing in 1961 have been grouped according to the subdivision of 1911. There are too few prior socioeconomic data to make a meaningful comparison with earlier periods.

The distributions across the whole territory of the factors under study, indicated in section 9.5, reveal a completely different situation at the two times (table 9.2), reflecting the profound and general modification in Italy's socioeconomic conditions, especially after the Second World War. At the beginning of the twentieth century a relative homogeneity of environmental factors existed among Italian provinces, at least for demographic quantities, as indicated by a compari-

TABLE 9.2. Environmental variables and consanguinity: mean, standard error (SE), and coefficient of variation (CV) of the Italian provinces in 1911 and 1961

| Environmental variables | 1911 | | | 1961 | | |
|---|---|---|---|---|---|---|
| | Mean ± | SE | CV% | Mean ± | SE | CV% |
| Live births | 32.1 ± | 0.59 | 14.4 | 17.3 ± | 0.52 | 23.5 |
| Deaths | 21.0 ± | 0.35 | 13.1 | 9.4 ± | 0.16 | 13.5 |
| Stillbirths | 41.7 ± | 0.78 | 14.6 | 21.0 ± | 0.99 | 36.7 |
| Infant mortality | 124.6 ± | 3.08 | 19.3 | 36.1 ± | 1.37 | 29.7 |
| Immigration | 22.4 ± | 2.09 | 72.8 | 33.9 ± | 1.61 | 37.0 |
| Internal emigration | 22.1 ± | 2.54 | 89.9 | 39.3 ± | 1.57 | 31.2 |
| External emigration | 4.1 ± | 0.49 | 94.1 | 0.8 ± | 0.12 | 128.7 |
| Ruralization | 593.6 ± | 17.20 | 22.6 | 336.8 ± | 17.66 | 41.0 |
| Industrialization | 270.1 ± | 11.79 | 34.1 | 362.8 ± | 13.08 | 28.2 |
| Illiteracy | 390.5 ± | 24.56 | 49.1 | 87.6 ± | 8.19 | 73.0 |
| Population density | 152.1 ± | 22.71 | 116.7 | 205.1 ± | 33.36 | 127.0 |
| First cousins | 21.0 ± | 1.81 | 67.4 | 10.6 ± | 1.60 | 117.5 |
| Second cousins | 15.4 ± | 1.09 | 51.9 | 8.4 ± | 0.68 | 59.0 |

son of the coefficients of variation in 1911 and 1961, with the exception of internal migration. On the contrary, in 1961, economic development seemed to have influenced differentially all demographic and some social factors, including the variation of consanguinity (especially for first cousins).

We can try to explain some aspects of the difference between the demographic and socioeconomic structure of Italy in 1911 and 1961 through an analysis of the changes in the relationships among the environmental factors (table 9.3). In fact, the correlations between the pairs of variables differ in intensity and/or in sign, when the two periods are compared.

*Birth rate* and *death rate* provide the most telling examples: their correlation was significant and positive in 1911, and highly significant but negative in 1961. In 1911 the death rate was associated with a condition of low social and cultural development (low internal mobility, high infant mortality, and illiteracy), while in 1961 it was directly associated with the factors that indicate socioeconomic development (high industrialization and high internal mobility). Therefore, the correlations of all factors that include the death rate will have the opposite sign when considered for the two different time periods.

Another aspect of the difference between the two periods is shown by the degree of association between *demographic and social factors* (livebirth, death, stillbirth rate, infant mortality) and *economic factors* (ruralization, industrialization): weak or absent in 1911, it is very strong in 1961. The high and negative correlation between illiteracy and economic factors that also existed in 1911 was an exception to this general trend.

A positive very strong correlation between *immigration and emigration* was detected in the provinces in 1911 ($r = 0.98$, $P \ll 0.001$); the correlation in 1961 ($r = 0.27$, $P = 0.034$) was still positive but much less significant. This behavior, which is the opposite of the expected negative relationship, was interpreted by demographers as being due to the peculiar type of migratory exchange relating only to the neighboring zones, so that immigration and emigration balance each other. Therefore, differences in the intensity of this relationship will be proportional to that part of migration involving short-distance displacements. Relaxation of this positive correlation in the course of time could indicate that socioeconomic evolution (and the resulting establishment of attraction poles) increased long-distance migration, thus diversifying the weight of the two migration components, immigration and emigration.

The migration rate is a particularly important parameter when variations in consanguineous marriage frequency need to be studied. In fact, the immigration of new individuals into a given area from outside or the emigration of the inhabitants of an area toward other zones regulate the number of related persons and influence (together with other cultural factors) the "marriage choice." Moreover, the average distance within which individuals moved can also affect the demography of an area.

The great socioeconomic changes that occurred over fifty years can justify our observation of different degrees of association between migration and other environmental factors measured in different time periods: 1911 and 1961. Yet, within each of the two periods considered, the territorial heterogeneity of economic development may also have determined different migratory behaviors, thereby later influencing consanguineous marriage frequency.

The results displayed in table 9.4 show that in 1911—that is, in a period of generally low economic development—economic factors, such as *ruralization* and *industrialization*, and some demographic or social factors, such as *livebirths, stillbirths,* and *infant mortality*, did not influence the frequency of first-cousin marriages, which, on the contrary, seem to be affected by the variation of *migration rates* (internal and external). Other types of factors different from those considered here (perhaps customs) increased the frequency of unions between close relatives, such as those between first cousins. Variations in levels of economic development, (*ruralization* and *industrialization*) and in ecological factors (*altitude* and *density*), which are highly correlated, seem to have been important in determining only the frequency of marriages between more distant relatives, which are less subject to family ties.

Fifty years later, when the degree of economic development diversified the provinces according to their demographic, industrial, and social structure, almost all the environmental variables are associated with the frequencies of marriages between first cousins and between second cousins, indifferently.

Differences between the two time periods are emphasized to a

TABLE 9.3. Italian provinces: correlation between pairs of environmental variables in 1911 and 1961

| Environmental variables | Live births | Deaths | Stillbirths | Infant mortality | Immigration |
|---|---|---|---|---|---|
| Deaths | | | | | |
| 1911 | 0.35** | | | | |
| 1961 | −0.72*** | | | | |
| Stillbirths | | | | | |
| 1911 | −0.08 | 0.27* | | | |
| 1961 | 0.69*** | −0.52*** | | | |
| Infant mortality | | | | | |
| 1911 | 0.64*** | 0.61*** | 0.20 | | |
| 1961 | 0.77*** | −0.46*** | 0.85*** | | |
| Immigration | | | | | |
| 1911 | 0.02 | −0.41*** | −0.04 | −0.19 | |
| 1961 | −0.69*** | 0.63*** | −0.64*** | −0.60*** | |
| Internal emigration | | | | | |
| 1911 | 0.12 | −0.42*** | −0.06 | −0.16 | 0.98*** |
| 1961 | −0.25 | 0.15 | −0.20 | −0.22 | 0.27* |
| External emigration | | | | | |
| 1911 | −0.15 | 0.30* | 0.18 | 0.17 | −0.51*** |
| 1961 | 0.40** | −0.31* | 0.57*** | 0.47*** | −0.42*** |
| Ruralization | | | | | |
| 1911 | 0.26* | −0.13 | 0.11 | 0.11 | −0.16 |
| 1961 | 0.28* | −0.40** | 0.59** | 0.44*** | −0.42*** |
| Industrialization | | | | | |
| 1911 | −0.22 | 0.07 | −0.16 | −0.12 | 0.21 |
| 1961 | −0.31* | 0.43*** | −0.59*** | −0.38** | 0.42*** |
| Illiteracy | | | | | |
| 1911 | 0.18 | 0.30* | 0.25 | 0.38** | −0.54*** |
| 1961 | 0.70*** | −0.71*** | 0.88** | 0.71*** | −0.68*** |
| Population density | | | | | |
| 1911 | −0.05 | 0.31** | 0.26* | 0.10 | 0.01 |
| 1961 | 0.29* | −0.05 | 0.08 | 0.15 | −0.03 |
| Altitude | | | | | |
| 1911 | −0.03 | 0.24 | 0.21 | 0.06 | −0.54*** |
| 1961 | 0.15 | −0.15 | 0.42*** | 0.43*** | −0.36** |

*Significantly different from zero at $P < 0.05$.
**Significantly different from zero at $P < 0.01$.
***Significantly different from zero at $P < 0.001$.

| Internal emigration | External emigration | Ruralization | Industrialization | Illiteracy | Population density |
|---|---|---|---|---|---|
| −0.56*** | | | | | |
| −0.22 | | | | | |
| −0.06 | 0.30** | | | | |
| 0.33** | 0.30** | | | | |
| 0.12 | −0.35** | −0.94*** | | | |
| −0.25 | −0.20 | −0.87*** | | | |
| −0.49*** | 0.63*** | 0.37** | −0.40** | | |
| −0.08 | 0.52*** | 0.69*** | −0.68*** | | |
| −0.02 | −0.15 | −0.53*** | 0.48*** | −0.09 | |
| −0.26* | −0.10 | −0.35** | 0.16 | −0.02 | |
| −0.52*** | 0.46*** | 0.50*** | −0.51*** | 0.29* | −0.32* |
| −0.16 | 0.34** | 0.38** | −0.15 | 0.35** | −0.32* |

TABLE 9.4. Italian provinces: correlation between environmental variables
and frequencies of first and second cousins in 1911 and 1961

| Environmental variables | First cousins | | Second cousins[a] | |
|---|---|---|---|---|
| | 1911 | 1961 | 1911 | 1961 |
| Live births | −0.07 | 0.62*** | −0.02 | 0.41** |
| Deaths | 0.36** | −0.45*** | 0.29* | −0.36** |
| Stillbirths | 0.04 | 0.81*** | −0.01 | 0.49*** |
| Infant mortality | 0.20 | 0.65*** | 0.03 | 0.48*** |
| Immigration | −0.57*** | −0.61*** | −0.52*** | −0.67*** |
| Internal emigration | −0.61*** | −0.27* | −0.54*** | −0.31* |
| External emigration | 0.64*** | 0.57*** | 0.33* | 0.43** |
| Ruralization | −0.07 | 0.42*** | 0.31* | 0.27* |
| Industrialization | 0.02 | −0.45*** | −0.37** | −0.13 |
| Illiteracy | 0.40** | 0.77*** | 0.06 | 0.42** |
| Population density | 0.01 | 0.01 | −0.35** | −0.36** |
| Altitude | 0.32* | 0.28* | 0.69*** | 0.72*** |

[a]Without Sicily.
*Significantly different from zero at $P = 0.05$.
**Significantly different from zero at $P = 0.01$.
***Significantly different from zero at $P = 0.001$.

greater extent by a principal component (PC) analysis (results not given in detail). Only 34% of the variation among provinces of the 12 environmental factors in 1911 is explained by the first component to which factors indicating the mobility of the population (*immigration* and *emigration* with the same sign) and *industry* (the "cause" of the movement) contribute most significantly. Fifty years later, demographic and social factors—*illiteracy, stillbirths, infant mortality, immigration*—explain 48% of the variation among the provinces.

A different approach can provide some information about the effect of the change of the environmental variables on the decrease of close consanguinity (marriages between first cousins) or more remote consanguinity (marriages between second cousins). The difference between mean values of the 11 variables (excluding *altitude*) in years 1961 and 1911, related to the initial values in 1911, for all the provinces was correlated with the time difference in frequencies of first cousins and of second cousins. The synthetic indices obtained from PC analysis on evolution rates of environmental variables (table 9.5) show that the first PC (and the third) is very highly correlated with

TABLE 9.5. Italian provinces: PCA on mean difference 1961–1911 on the environmental variables

| Rank of variables | PC | | |
|---|---|---|---|
| | *I* | *II* | *III* |
| 1 | Internal emigration | Ruralization | Population density |
| 2 | Illiteracy | External emigration | Industrialization |
| 3 | Stillbirths | Industrialization | Deaths |
| 4 | Live births | Population density | Infant mortality |
| 5 | Immigration | Immigration | Internal emigration |
| 6 | Deaths | Infant mortality | Immigration |
| 7 | Infant mortality | Illiteracy | Stillbirths |
| 8 | Ruralization | Internal emigration | Live births |
| 9 | Industrialization | Deaths | Ruralization |
| 10 | Population density | Live births | External emigration |
| 11 | External emigration | Stillbirths | Illiteracy |
| % of total contribution per PC | 47.2 | 14.9 | 12.1 |
| Correlation with time change in cousins | | | |
| First cousins | 0.806*** | −0.049 | 0.507*** |
| Second cousins | 0.199 | −0.017 | 0.175 |

***Significantly different from zero at $P = 0.001$.

decrease of first-cousin marriages only. Changes in time of second-cousins marriages do not seem to parallel changes of this type of factors. The simple correlations between each environmental variable with both types of cousins (table 9.6) support these findings. Interestingly, the factors more associated with decrease of first-cousin marriages (*illiteracy, internal emigration, stillbirths*) are those that contribute more to first PC in the analysis describing change in de-

TABLE 9.6. Italian provinces: correlation between changes of environmental variables and changes of consanguinity over time (1961–1911)

| Environmental variables | First cousins | Second cousins[a] |
|---|---|---|
| Live births | 0.61*** | 0.06 |
| Deaths | −0.68*** | −0.28* |
| Stillbirths | 0.77*** | 0.16 |
| Infant mortality | 0.42*** | −0.02 |
| Immigration | 0.50*** | 0.10 |
| Internal emigration | 0.77*** | 0.24 |
| External emigration | −0.14 | −0.10 |
| Ruralization | 0.59*** | 0.06 |
| Industrialization | −0.29* | 0.11 |
| Illiteracy | 0.82*** | 0.20 |
| Population density | 0.23 | 0.01 |

[a]Without Sicily.
*Significantly different from zero at $P = 0.05$.
**Significantly different from zero at $P = 0.01$.
***Significantly different from zero at $P = 0.001$.

mographic and socioeconomic structure of Italy. Strictly economic change, as the percentage of industrial occupation, seems to be less important.

Behind all the demographic changes in birth and death rates are changes in education and hygiene. Behind all the changes in migration are changes in opportunities and general economic development. It is to these changes that we must look for to understand the trend of consanguinity. Naturally, in some parts of the country, especially the north, there has been a major increase in industrialization, which has contributed to the economic improvement in a major way. But such change has been lacking in the south, especially in the first half of the twentieth century. Apart from this aspect, which was probably not of major importance regarding the change in consanguinity, the process of sociocultural change has followed similar paths in the north and south, as shown by the correlation between the first PCs of 1911 and 1961, which summarize the major socioeconomic patterns observable at the two dates. Figure 9.11 shows that both first PCs are highly correlated with the patterns of consanguinity change (measured by variation of first cousin marriages) that we have described in section 9.4.

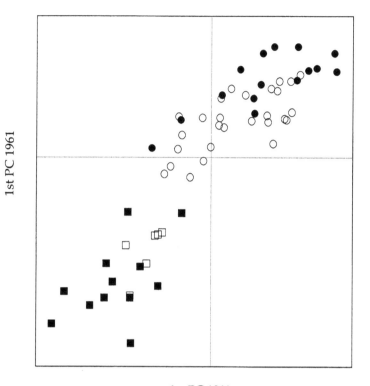

1st PC 1961

1st PC 1911

● 1 North-West area
○ 2 North-East and Central area
■ 3 South and Sardinia
□ 4 Sicily

FIGURE 9.11. Correlation between the first PC' on socioeconomic variables for Italian provinces in 1911 and in 1961 ($r = 0.89$, $P < 0.001$). Provinces are grouped according to the patterns of consanguinity change represented in Figure 9.7.

## 9.8 AN ATTEMPT AT A GENERAL SYNTHESIS

The Vatican data are especially informative for the first half of the twentieth century, during which time consanguinity has, in general, decreased after a first small increase. This is also the period during which major socioeconomic changes have taken place. But the nineteenth century saw first a slow, then a faster increase of consanguinity. Unfortunately, there are few data that show the whole process from the beginning of the nineteenth century to the end of the twentieth century. Figure 9.12 comes closest to this need, with comparative data for a very long period for two northern dioceses, Piacenza and Reggio Emilia, and for three large areas in the south, Sardinia, Sicily, and the Aeolian Islands, which represent various southern situations. There are two very distinct phases: first, an increase, at first slow, then faster, which ends around 1915–1925 when a peak of consanguinity is reached, at a similar time for all Italian regions, and, second, a rapid descent begins.

We have already considered on earlier occasions some of the possible explanations of the first phase, which are, in order of time: (1) changes of laws of inheritance; (2) decreased influence of the Roman Catholic Church; and (3) increase in relative abundance of cousins due to increase in population size, which increases disproportionately the abundance of relatives, though with a delay of one generation for first cousins and two for second cousins.

It is possible and even likely that all these factors play a role, but it is difficult to evaluate their relative importance. It would seem reasonable to infer that population growth plays a major role in the second part of the ascent phase, in which consanguinity increases faster. It is not so much the rate of increment of a population that plays a role in increasing the relative abundance of cousins, perhaps, but the rate at which the increment rate changes, and this is higher in the south, as can be seen in the curves of the demographic transition. This may explain qualitatively, at least, why the southern dioceses show a faster growth of consanguinity than the northern ones and reach higher consanguinity values at peak time. While the peak times are similar in all regions, the peak $\alpha$ values vary from 1.3‰ in Reggio Emilia, 2.2‰ in Piacenza, 2.6‰ in Sardinia, 4.9‰ in Sicily, to above 12‰ in the Aeolian Islands.

The second phase of the phenomenon, the decrease of consanguinity, must be the same as that called "the breakdown of isolates" by Dahlberg. He described it in Sweden, where it began in the second half of the nineteenth century. It is essentially due to an increase in

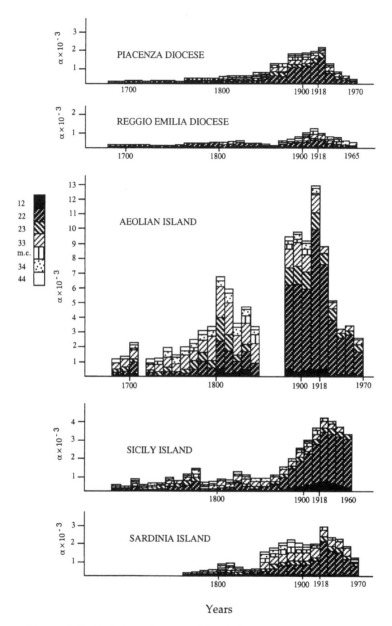

Years

FIGURE 9.12. Evolution of consanguinity during three centuries in five areas of Italy.

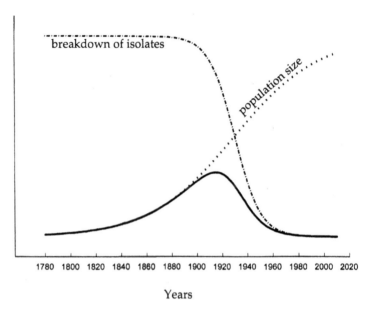

Years

FIGURE 9.13. A model to explain the peculiar trend of consanguinity in Italy (see text). We wish to acknowledge the help of A. Gimelfarb in drawing this figure.

individual mobility, tied to increased means and speed of transportation, and increased opportunities of work in specific industrial areas, favoring relocation of workers. It was definitely faster than the ascent, by a factor a little less than 3. It is most likely that the same phenomenon is occurring in our country, with a lag of perhaps half a century.

Figure 9.13 is an attempt at a simple, semiquantitative explanation of the whole process. The curves assumed to generate the phenomenon are two logistic curves, an ascending one that describes the trend of increasing population and a descending one for the breakdown of isolates. In Italy this phenomenon occurred later than in northern Europe, and was assumed to reach the 50% point shortly after 1925. The two logistics have been given slopes that differ as 1:3 to approximate the ratio of the slopes of the initial increase and the later decrease in consanguinity. No attempt was made at fitting the real data, which are too irregular to be accurately fitted. The solid curve is the product of the two logistics and is expected to describe approximately the consanguinity trend in time. Its peak approximates those observed in figure 9.12, which vary between 1915 and 1925, being slightly later in the south. The difference in the peak consanguinity values between north and south agrees with the lower growth rate of the northern populations, which we have described in section 9.6, and the major effect of demographic variables on consanguinity, which we have investigated in section 9.7. Our analysis in section 9.7 referred mostly to the decreasing phase, but the demographic transition study of section 9.5 include data of the relevant period in the nineteenth century. This observation supports the role of population growth on the increase of consanguinity, without denying a possible partial role of the other explanations.

The difference in the time at which the peak appears seems relatively small in the various parts of the country, indicating that the breakdown of isolates took place at similar times all over Italy and was a national rather than a local phenomenon. The descending curve would place its beginning at the turn of the century. There are tenuous indications that it was a little later in the south, but the nature of the phenomenon and a more precise analysis would demand a socioeconomic research that could hardly be part of this endeavor.

CHAPTER 10

# Geography of Demes in Italy

### 10.1 POPULATION SAMPLING

Populations, demes, and random mating are recurrent terms in the population genetics literature. This is a good opportunity for clarifying these concepts and sharpening their meaning.

Population has a precise statistical definition: it is the source of the samples of individuals that we study. It is important that it is defined carefully enough at the beginning so that if we take another sample from the same population, we can expect it to give the same statistical conclusions obtained from the first. But populations may change over time, and so we can expect only an approximate correspondence between independent samples at different times. Moreover, we may expect to obtain the same statistical conclusions from them, strictly speaking, only if the samples are random.

Rules of random sampling are well established but very rarely followed. Taking a random sample of individuals for genetic studies may pose special difficulties. Strictly speaking, we would want a list of the people forming a population and would choose randomly from it. This would require a census of the population, making it possible to numerically code all individuals and use random numbers to choose those who will form the sample (or other equivalent methods). But this would demand obtaining samples from them at their home or requesting them to come to special locations, a very difficult task. It is very rarely done. One often needs a blood sample from each individual, and some are likely to refuse: does this introduce a bias? Fortunately, as much as it is possible to judge, in practice, this does not seem a likely source of error, but there are situations in which an unwanted, hidden selection may occur. In the sampling of a population for genetic purposes we often do not choose individuals by the phenotype of the genes we study, and this is a help to avoid sampling biases—barring exceptional circumstances. When volunteers are requested to join by an appeal through media or other special sources, there may be other unexpected biases. Repeated sampling by different sampling methods or in different strata and testing the difference between independent samples may give some guarantee against biases.

A common problem when choosing phenotypes represented by specific diseases is that the definition of the disease is not as sharp as we would like it to be. Frequently blood samples for ethnic analysis are collected from hospitals even though the specific diseases affecting them will not be considered in the analysis. Sampling in hospitals requires much care in the choice of hospital departments, and the same is true for sampling from schools. In these and other similar cases socioeconomic stratification of the population generates special sampling problems, because different socioeconomic strata vary in ethnic and geographic origin of their members. These modes of collection generates special, not only statistical, but also bioethical problems. Genetic investigations pose strong constraints of confidentiality and anonymity with respect to individuals and groups to be kept continuously in mind.

But bias may arise in many ways, in particular, as a function of the places where sample collections are made. If samples are limited to cities, as is often the case, the actual geographic area represented by each city is usually a vast and confused hinterland. Large cities and, in particular, capitals receive much immigration from wider areas, which have changed with time. For smaller cities the hinterland is narrower, with fewer additions from a wider area. Therefore, trees of populations collected from different places show large cities or capitals as the "oldest" part of the tree. But this is erroneous: the tree of the geographic areas from which samples originate will usually have at its top those that received the widest collection of immigrants (for a clear discussion and example of this source of error see the analysis of Taiwan surnames by Chen and Cavalli-Sforza [1983], and also Piazza et al. [1987]). Thus, a tree of major cities loses much of its historical meaning and acquires new, unwanted ones.

"Stratified" sampling increases the accuracy of sample representativeness. A useful stratification program is that of choosing the number of individuals to be collected in a given geographic area proportionately to the number of individuals living in it. This criterion is especially good for a description of the geography of the population at present time, and may be necessary for research in epidemiological genetics. Because it is inevitably influenced by the attraction large cities have had for immigrants from large areas, as mentioned above, it is not the best for historical reconstruction. From this point of view it is better to stratify by geographic subarea, dividing the total area in subareas of approximately equal size that have precise geographic, ecological, socioeconomic, and hopefully also historical meaning. About equal or similar numbers of individuals are chosen from each subarea. This may be called geographic- rather than population-stratified sampling. When they are available, one can use additional criteria that make it possible to concentrate on the settlers of the periods of special interest.

Where surnames originated long ago, they can help in identifying older settlers (though not necessarily the oldest, who may have largely disappeared or moved elsewhere in earlier times). In Italy a great number of surnames are rare and tend to be concentrated in a small area, which is likely to be the area of origin of present inhabitants at some time earlier than the date of origin of the surname.

A current project of sampling the Italian population for genetic investigations uses the following approach. The general strategy uses surnames from sources such as telephone books (where telephones are nearly universal), electricity bills, or other population lists, such as voter lists. One studies the geographic distribution of households identified by surnames in geographic areas defined by the smallest administrative unit available. In Italy this is the *comune*, to be defined more specifically later. This unit is usually too small or, for large cities, too big, to become the minimal subpopulation unit employed in an analysis of genetic heterogeneity, and usually wider areas that include a suitable number of the smallest administrative units are chosen, on the basis of specific geographic, ecological, and/ or historical interest. These areas serve the purpose of identifying subpopulations corresponding to the minimal geographic units according to which the general population will be divided in a final heterogeneity analysis. Many surnames tend to cluster in characteristic ways in one administrative unit or a few adjacent units forming these subareas.

Such surnames are selected, and a list of them is passed to institutions collecting blood for donation or tissues for transplant with the request to collect blood samples from one (or more, a number to be specified) donor for each surname. The blood samples returned to the genetics laboratory must carry only the indication of the area of origin designed to represent the minimal geographic unit, on the basis of which a subpopulation will be identified in results to be made public. The identity of the blood donor is not made available to the genetic laboratory, and is replaced by a new identification code known only to the blood donor organization. Results of the individual genetic tests are not made known to the donor organization. Thus, no connection is established between the donor and the genetic tests done on his/her blood sample. In case this is of interest to or desired by the blood donor, it is always possible to reconstruct the connection between the donor and the test results, if the donor organization keeps the knowledge of the two identity codes it has established. These precautions are part of the general ethical principles that are necessary in the genetic study of populations.

The introduction of surnames is useful for identification of ancestral contributions. It helps in selecting direct descendants of the "founders," and allowing comparisons of the gene frequencies and ethnic origins of the various subareas. Ancestral contributions thus selected are, strictly speaking, valid only for the patrilinear side, but there is a calculable degree of correlation for the matrilinear side, if mtDNA markers are tested. These "founders" are a least as old as those at the time of origin of the surnames. In Catholic countries most surnames go back at least sixteen generations. It is possible that the selection of relatively rare surnames introduces some bias in this procedure, but it is unlikely to be a serious one, given that only a small minority of surnames are truly very common.

The use of geographic-stratified sampling may be of advantage for genetic purposes, such as the study of genetic disease, by allowing one to sample population founders and compare them genetically with patients of specific diseases (Siniscalco et al. 1999). For purposes of genetic epidemiology it is of interest to evaluate the correlation of patients with these diseases with population-stratified samples.

The frequencies of the sample of one type can be predicted from those of the other if there is knowledge of migration rates in the general area. This can be derived from demographic sources such as parish books by methods described by Wijsmann et al. ( 1984).

## 10.2 RANDOM MATING, MENDELIAN POPULATIONS, AND DEMES

Genetic analysis is simplified if the population being investigated is under random mating. A population is usually chosen for study on the basis of geography, ecology, or history. There are many possible causes of deviation from random mating. There is always some socioeconomic stratification, which is hard to evaluate accurately. Moreover, there is always another major source of deviation from randomness of mating, represented by the geographic distance between birthplaces or, more importantly, residences of mates. With very few exceptions, marriages take place between individuals born at very short distance. This automatically generates a strong "isolation by distance," which is most easily observed by testing the strong positive correlation, existing in practically every culture, between the genetic distance of individuals or groups and the geographic distance of their birth or residence places (see a summary by Wijsman and Cavalli-Sforza [1984], and correlation graphs for many genes in Cavalli-Sforza et al. [1994], complete edition, cloth bound). In the following we summarize population genetic models and their applications to observed data, which is largely taken from *The Genetics of Human Populations* by Cavalli-Sforza and Bodmer (1971, 1999), where fuller details can be found.

There is a classic genetic way of testing randomness of mating, the well-known Hardy-Weinberg (HW) distribution of genotypes. For genes for which at least two distinguishable alternative forms of a gene (alleles) are known, the standard HW rule predicts that homozygotes for an allele (individuals who have inherited the same allele from both parents) are expected to have frequency equal to the square of the frequency of that allele. Heterozygotes for two alleles (who inherited different alleles from their parents) are twice the product of the frequencies of the two alleles. The equilibrium condition at which the HW rule is satisfied is reached in one generation, and, at most, two if gene frequencies happen to be at the beginning different in the two sexes. It is thus verified in practice even if a population has been subjected to a relatively recent admixture of populations having different frequencies of those alleles. Statistical data from a great number of populations and genes, such as those collected in Mourant et al. (1976), show that almost all human population samples examined satisfy the HW rule, unless they are very large (as happens for samples of blood donors that originate from a very wide area), or the gene is poorly known, or the alleles not easily distinguished, or there is strong selection, assortment of mates, or inbreeding.

Very large populations cannot be considered the result of random mating, mostly because there is isolation by distance, and most genes show some difference in gene frequencies for geographically distant populations. In the integral edition of *History and Geography of Human Genes* (Cavalli-Sforza et al. 1994) this relation is shown in diagrams accompanying the geographic maps of allele frequencies for many genes and continents. This source of genetic heterogeneity can be tested by plotting the difference of gene frequencies among two populations as a function of their geographic distance. Half the square of the difference of the frequencies of an allele in two populations, $p_1$ and $p_2$, divided by the average frequency of that allele, $p$, and by $1 - p$, is called $F_{ST}$ (also Wahlund variance). $F_{ST}$ between two populations is a measure of genetic distance between them, and it increases linearly, on average, with geographic distance between the two populations. There is considerable genetic variation between pairs of populations examined and therefore linearity of genetic versus geographic distance shows clearly only when the number of samples of gene frequencies is large and robust calculations of averages are employed.

At high geographic distances the $F_{ST}$ measure, which can never be greater than 1, tends to an asymptote, which is somewhat different in different continents. Two populations that have equal gene frequencies for an allele have zero $F_{ST}$, and this is usually nearly true for geographically close populations. The average $F_{ST}$ of all world populations in humans for single alleles is around 0.13. But mixtures of individuals from populations including subgroups that do not mate randomly and have substantial heterogeneity of gene frequency of an allele, that is, an $F_{ST}$ greater than zero, have a deficit of heterozygotes and thus do not satisfy HW. The same is true of a population that has undergone very recent admixture, and there has been no time for establishing HW equilibrium. It is also true if a population is subject to another cause of deviation from random mating, *inbreeding*. The deviation of the observed frequency of heterozygotes from their HW expected frequency allows us to measure its average inbreeding coefficient (see section 1.5), but there is no guarantee that it is the only cause of deficit of heterozygotes.

Another cause of deviation from HW is *selection*, if the frequency of genotypes is measured after selection took place. For instance, alleles with strong pathogenic effects, like those for sickle cell anemia or thalassemia, do not satisfy HW. Pathogenic alleles for these genes are recessive and, especially in countries with poor sanitary conditions, most homozygotes for these genes die before puberty. A high frequency for one or the other of these lethal genes is caused in populations that have been affected by malaria for a long time, because heterozygotes are resistant to this disease. Here heterozygotes have a frequency satisfying HW expectation if the sample is tested at birth or on very young children, but higher than that expected by HW among adults. Even though selection is very strong in this case, and the difference in selective values of normal homozygotes and heterozygotes may be as high as 10% in the presence of malaria, fairly large samples are required for demonstration of a deviation from HW.

Another condition causing deviation from HW is *assortative mating (a.m.)*. Positive a.m. is observed in humans for some observable quantitative traits, for example, tall choosing tall and short choosing short mates (with a correlation coefficient around 0.30—some of it secondary to assortment for socioeconomic or ethnic reasons). Assortment for psychological and socioeconomic traits is even more important than that for stature (it is around 0.5 both for IQ and for socioeconomic conditions). Negative assortment may be observed, though probably very rarely. Strictly speaking, a.m. should be kept distinct from sexual selection, although a.m. may cause it. An example of sexual selection would be the attraction exercised by handsome people, male or female. An example of positive a.m. probably causing sexual selection is the correlation between spouses for height. Such assortment may cause selection against the extremes, which was noted, for instance, for stature and for chest girth (Cavalli-Sforza and Bodmer 1971, 1999). But selection against extremes may have other causes.

The term "Mendelian populations" was introduced by Dobzhansky (1951) to indicate populations under random mating. Ideally, for most purposes of historical and evolutionary study, one would like to take samples of genes and populations that satisfy this condition. But no real population is effectively under true random mating. We have given examples of assortative mating, which are fairly general for the somatic traits we mentioned. On the other hand, the validity of the HW rule for almost all genes supports the idea that mating is almost universally random. Is there a contradiction between these findings? If a gene satisfies HW, others are not necessarily expected to satisfy it. But the statistical test for HW also depends inevitably on the number of individuals tested for HW validity. The chance of detecting statistically significant deviations depends on that number and on the magnitude of the deviation. In general, the sensitivity of the test is low and the number of individuals necessary to detect a small deviation is very high, irrespective of whether it is caused by inbreeding, isolation by distance or other causes of nonrandom mating, or natural selection. One cannot usually distinguish between the hypothesis that a test for HW deviation is negative because not enough individuals have been tested, or because mating is truly random. The demonstration of assortative mating for quantitative traits is also affected by the nature of genetic control of such traits. They often are determined by many genes (i.e., they are polygenic), and the contribution of each gene to the overall value of the trait may be so small, that the probability of finding a deviation from HW for one component of the polygenes, because of assortative mating, is negligible. In practice, the chance of detecting natural selection by deviation from HW expectation is also quite low.

The test of random mating by HW is very useful for a certain number of genetic problems, but only moderately or not at all for taxonomic purposes. "Deme," a word suggested by two botanists, Gilmour and Gregor (1939), and one coming slowly into common usage, is shorter than "Mendelian populations" and has less constraints. It was originally defined as "any assemblage of taxonomic closely related individuals." The authors were aware that most "breeding communities," another expression they use for the same purpose, are rarely fully isolated and that "when more experimental work will be done on this point it should be possible to devise a method of expressing degree of isolation quantitatively." This was certainly prophetic, since the measure of drift is the quantity $Nm$, where $N$ is population size and $m$ the migration rate, or the deviation from complete isolation. In the following we will use the word "deme" and show that it is quite flexible. Gilmour and Gregor indicated that the word can be specialized to cover different aspects and suggest "gamodeme" for reference to breeding, "topodeme" to geography, and "ecodeme" to ecology, but the specification is rarely necessary or even easy.

The regularity of increase of genetic distance with geographic distance of population pairs, when studied at the world level or over large regions like continents or subcontinents, does not allow one to set a priori a "threshold geographic distance" at which two populations can be considered reciprocally isolated genetically and labeled different demes. Nor is there any indication that some particular genetic distance can be chosen as threshold for the same purpose. On average, the increase of genetic distance with geographic distance is continuous and regular. Even so, it is possible that there exists a clear genetic heterogeneity even among populations living in the same area, when the reciprocal isolation is due to reasons other than geographic distance (e.g., rarity or total lack of cross-marriages between the groups for religious, cultural, or socioeconomic motivations) and has lasted for long periods of time. Such populations may well be considered as different demes on the basis of these genetic and demographic criteria.

It is useful to keep in mind that with random mating and absence of natural selection, in populations that originated from the same initial one and remained partially or totally isolated among themselves for enough generations, the deviation from HW shown by the frequency of heterozygotes in an artificial mixture of these populations is equal to the $F_{ST}$ variance of gene frequencies between the populations (Wahlund's formula). Under these conditions the two approaches give exactly the same results. The presence of other evolutionary factors may destroy this equivalence.

Deviation from randomness in mating could be observed more easily by studying the joint distribution of alleles in two or more separate genes by "linkage disequilibrium (LD) tests," which measure the nonrandomness of the distribution of alleles at different loci, having a different position on the chromosome. But this kind of deviation depends on population history in a complicated way. Alleles at two loci are not expected to be distributed randomly, although each locus follows the HW distribution, if they originated from a recent admixture of populations having different allele frequencies at different genes. When deviation from HW is not maintained by natural selection, but is established, for instance, by admixture of populations with different gene frequencies, linkage disequilibrium between the loci will disappear after the admixture, the more rapidly, the greater the recombination frequency between the loci, but it may take many generations if the differences in gene frequencies of the two populations were of some magnitude at the beginning, or if the two loci are close on the same chromosome. The rate of disappearance is maximum for a pair of genes located on different chromosomes or remote from each other on the same chromosome. Only genes that are very close on the same chromosome or whose combinations are under strong selection maintain observable LD for a great number of generations. For an elementary treatment see Cavalli-Sforza and Bodmer (1971, 1999).

Today, the study of linkage disequilibrium is at the center of genetic attention, mostly for the opportunities it offers to detect the location and nature of genes determining disease. The major difficulty is the extreme variation of linkage disequilibrium observed in different parts of the genetic map. This finding is of considerable interest also for the study of the nature of crossing over and other mechanisms of recombination and for badly needed new approaches to understanding natural selection. It is also very useful for reconstructing aspects of human evolution. It offered a strong confirmation of the hypothesis that the first demographic explosion of modern humans took place in Africa, that it was followed by a second explosion after the extension to Asia, and from here to the other continents.

## 10.3 COMPARING GENETIC AND DEMOGRAPHIC
## APPROACHES TO THE STUDY OF DEMES

The main interests of defining demes come from the desire to measure drift, to understand its contribution to evolution and its relative role compared with that of natural selection, and to use the vagaries of human population structure to detect more easily the contributions of specific genes to genetic diseases and other traits, especially in clearly polygenic situations. The questions to be asked are Is it necessary to define demes and estimate their sizes? Is it really useful? The answers depend on the nature of the population being considered and the questions to be asked. The great majority of human populations are not closed entities, that is, they are not true isolates that remained isolated for a long time without any exchange with other populations. One such true isolate is Samaritans, whose isolation may go back as far as the religious event that defined their origin about 3800 years ago (Bonné-Tamir 1980). In such cases it is not interesting to use the frequency of cousin matings, à la Dahlberg, or the HW deviation to estimate their size, because this can usually be evaluated directly from demographic data. If anything, it is more interesting to reverse the question and ask if the frequency of cousin matings or the HW deviation corresponds to the expected one, and thus test if certain consanguineous matings are favored or boycotted. If the test is satisfied, both the demographic isolate size and the frequency of cousin matings can be used to evaluate drift. Unfortunately, this is not as easy as one might hope, because population size did not remain constant, but may have fluctuated considerably in mostly unknown ways, and it is therefore difficult to calculate expectations from existing theory. It may be more satisfactory, but most laborious, to do it by simulations, and there remains an inevitable uncertainty because of the usual lack of adequate historical information on the fluctuations of population size (or other unknown quantities).

   In the great majority of cases, however, populations undergo genetic exchanges, and a major factor becomes isolation by distance. The relationship between migration and geographic distance has been examined by many, including the authors of this book. The mathematical treatment at the basis of theories by G. Malecot, S. Wright, and M. Kimura uses the usual diffusion equations (the equivalent of Brownian motion), generating an expectation of a Gaussian relationship between the probability of migration to distance $x$ and the distance itself. But this relationship very seldom fits observations. The logarithm of the probability falls less rapidly than with the square of $x$, as expected for the Gaussian function. It sometimes falls with $x$ (generating a simple negative exponential relationship—e.g., Cavalli-Sforza and Hewlett [1982]) and other times even with its square root (see Cavalli-Sforza 1962). The use for humans of a variety of methods of transportation may generate a multimodal distribution (Cavalli-Sforza et al. 1966), making recourse to heavy numerical analysis necessary.

It is therefore not enough to know the effective population size $N_e$, to estimate the amount of random genetic drift expected in a population. The necessary quantity is $Nm$ (where $N$ should really be taken as effective population size $N_e$). There are methods that can estimate $Nm$ directly, and there is no need to estimate separately the two quantities $N$ and $m$, but most demographic approaches will give rise to separate estimates. If we want to compare the expected amount of drift with the observed one, we must evaluate the first by a *demographic* approach and the second by a *genetic* approach. Remember that it is necessary to correct the total census size $N$, estimated demographically, by limiting it to the number of effectively reproducing individuals, or the so-called *effective population size $N_e$*. We will not enter into the details of the variety of effective population sizes suggested in the literature, such as that related to the variation of gene frequencies among populations, a potentially different one related to inbreeding, and others. We limit our interest here to the first. In fact, we have seen that a direct estimate of inbreeding from consanguinity or other genealogical methods is difficult and likely to lead to serious underestimates, while that of gene frequencies is on stronger grounds. A simulation given in Cavalli-Sforza and Bodmer [1971, 1999, pages 419–421] shows that for human populations a simple estimate of $N_e$ from census $N$ values is approximately one-third of the census size, that is, that part of a population which is active in giving birth to the next generation. The value of 1/3 may be somewhat different as a function of the demographic parameters of the population; a general formula was given by Nei and Imaizumi (1966), but in human populations the necessary correction is small.

The estimation of $m$, the migration component, is limited to in-migration into the population unit whose $N$ is estimated, and is the proportion of immigrants per generation. This is quite different from the usual demographic estimates of migration, which are for shorter periods.

The comparison between observed and expected drift can be done in much more subtle and precise ways, but one simple approach is through the following approximate formula, tying the quantity $Nm$, the product of population size times the proportion of immigrants per generation estimated demographically, to the variance of gene frequencies among populations:

$$F_{ST} = \frac{1}{1 + KNm} \qquad (10.1)$$

with

$K = 1$ for the Y chromosome (NRY or nonrecombining portion
        of it), and for mtDNA
$K = 3$ for the X chromosome
$K = 4$ for autosomal genes

With (10.1) one can generate an estimate of the expected value of $F_{ST}$, using demographic data from which $Nm$ has been calculated, and compare it with the $F_{ST}$ value observed from gene frequency data. This is, of course, what was done in chapters 5 and 6.

Usually $Nm$ is much larger than 1, and therefore $F_{ST}$ is simply the reciprocal of $Nm$. Equation 10.1 ignores another important source of variation, the mutation rate $\mu$. Instead of $m$ a more correct expectation is $1 - (1 - m) \times (1 - \mu) = m + \mu$ approximately, but $\mu$ is much smaller than $m$ for nonisolated populations, and, in practice, the contribution of $\mu$ can be neglected except in very special situations.

In earlier chapters we have seen that in the area studied in the great detail, the upper Parma valley, there is, within the approximation possible in this study, reasonable agreement between estimates of observed and expected drift. This means that most of the genetic variation we observe is due to random genetic drift. In practice, this is true of the whole Parma valley, because in the lower part no variation due to drift is expected with the samples investigated, or is observed. This does not diminish the importance of natural selection, which remains the only major force generating adaptation to local environments, and the consequent differentiation of groups living in different ecological niches. Natural selection, however, does not affect all genes all the time—on the contrary, the great claim by Kimura in his 1968 paper was that, in molecular evolution, most genes are free from natural selection most of the time. At the more general level, the relative role of natural selection and drift has not been evaluated quantitatively and will vary with organisms, times, and environments. But in areas comparable to that of the Parma valley it is only drift that is of sufficient magnitude to be observed, at least for the great majority of genes.

Equation 10.1 can give us considerable freedom in the choice of a deme, but we must still choose it in a range defined by clear lower and upper bounds. The lower bound must be large enough and the upper bound small enough that they include conditions of "random mating." Randomness must be with regard to the inherited characters studied, though not with respect to age, socioeconomic status, or distance between birth or residence places of spouses. The last of these three constraints is isolation by distance, which is under control if both $N$ and $m$ are known. Ideally one would like to know a simple relationship between $N$ and $m$ for various sizes of demes (i.e., for various $N$), for then one may choose deme size at will. We will suggest such a relationship in the next section.

Jorde (1980) reviewed $F_{ST}$ magnitude in human populations and showed that different subdivisions of populations affect $F_{ST}$ values. Cavalli-Sforza and Feldman (1990) studied genetic data from the upper Parma valley from 37 parishes (already analyzed in chapter 5) at three levels of subdivision, calculating $F_{ST}$ among the 37 parishes and also after grouping these into 11 neighborhoods (subareas) and again after grouping the neighborhood in four communes (a standard administrative unit). The average number of inhabitants per parish in 1951 was 358.1. The $F_{ST}$ values decreased regularly with increasing size of the deme: parish → subarea → commune, approximately like the reciprocal of the deme size. Table 10.1 shows the fit of a linear

TABLE 10.1. $F_{ST}$ values ± standard errors (SE) at three levels of
population clustering of the upper Parma valley

| Number of clusters | Number of villages per cluster | $F_{ST}$ ± SE | Expected $F_{ST}$ for linearity |
|---|---|---|---|
| 37 parishes | 1 | 0.0261 ± 0.0059 | 0.02568 |
| 11 subareas | 3.36 | 0.0080 ± 0.0019 | 0.00804 |
| 4 communes | 9.25 | 0.0032 ± 0.0013 | 0.00319 |

*Source.* Cavalli-Sforza and Feldman (1990).

regression between log $F_{ST}$ and log of the numbers of parishes per cluster.

Random mating is not the only condition for choosing a population unit as deme. The general area in which we examine genetic heterogeneity by quantities such as $F_{ST}$ among demes may be large. The larger it is, the more genetic heterogeneity we may find, giving more power to the analysis. But there must be no strong heterogeneity of environmental conditions in the general area studied, or else part of the genetic variation among demes might be due to local differences in natural selection. We could hardly expect the absence of environmental heterogeneity when we examine variation over the whole world, unless only truly selectively neutral genes are included (see Cavalli-Sforza et al. 1994). Once these two conditions are satisfied (random mating within demes and absence of excessive environmental heterogeneity), we may be free of choosing as demes any type of population clusters. We will need, however, estimates of $N$ and $m$ for each deme to compare observed and expected genetic variation under drift alone.

It is not necessary to study various levels of population clustering for testing the agreement of demographic and genetic estimates of drift. One can simplify the test by estimating $Nm$ for a single deme type, but it remains necessary to evaluate, for the arbitrary average deme chosen, the size of the population in it as well as the average proportion of immigrants per generation (i.e., spouses from other neighboring population units). One calculates for the $i$th deme (Cavalli Sforza 1986)

$$Nm_i = \Sigma m_{ij} N_j \tag{10.2}$$

where $m_{ij}$ is the probability that an individual moves from village $j$ (of census size $N_j$) to village $i$ for marrying an individual of that village. This value can be averaged over all villages and supplies an estimate of $Nm$ for the whole area and the population clustering chosen. The proportion of endogamous marriages (both marriage partners from the same village) is $m_{ii}$ in village $i$. The $m_{ij}$ elements come from the migration matrix and all rows sum to 1.

The demographic analysis reported by Cavalli-Sforza (1986) for the Biaka (a group of African Pygmies living in the southwest of the Central African Republic) offers real data from another human population. A statistical analysis showed that the smallest population cluster, the camp, is made of 32.6 people per camp, living in 9.6 huts per camp. But the camp is highly movable and even though its composition has some degree of permanence during short periods, both composition and size vary with the time of the year (see also Bahuchet 1979). Camps (*lango* in local language) are made of families who are related mostly, but not exclusively, patrilineally, and easily accept guests from other bands or tribes (including western visitors).

A Pygmy camp is clearly not a random mating population. This consideration and the very small average size make camps inappropriate as measure of the deme. An analysis by Wobst (1974) showed that the minimum population size to avoid excessive inbreeding is of the order of 200–400 individuals. In some historical cases a population was for some time well below this number. Certain exceptional populations, like those of Pitcairn Island, Tristan da Cunha (Roberts 1968), and Pingelap (Morton et al. 1972) (see also Bodmer and Cavalli-Sforza 1976, pp. 381–385), went through bottlenecks even smaller than that of a Pygmy camp, without apparent genetic damage, but at least the first two populations were started by individuals of relatively remote ethnic origins, which assured a stronger genetic variation at the very start. A population size of this magnitude has been tolerated among Eskimos (Laughlin 1950) without loss of fertility, by exceptional measures that allowed one to avoid marriage for consanguinity greater than third cousins.

Among Pygmies, the next higher cluster above the camp is the hunting band, which is said to be a loose patrilineally exogamous unit and is certainly larger than the camp, but is not as sharp an entity as one might like and is not easy to pin down in practical surveys. Its size may be around 100. It seemed reasonable to choose other, larger units as potential "demes": in particular, all the Pygmy camps gravitating around a single farmers' village (or a tight cluster of farming villages) were considered demes. This group tends to be relatively constant from year to year. Many Pygmies spend only the dry season, the winter, in the village where they work for the farmers, and tend to go hunting in the forest at other times of the year.

It was not always easy to ascertain the degree of relationship among Pygmies from different camps living around a relatively large farmers' village or village cluster, but they seem to almost always speak the same dialect. Altogether, nine farmers' areas in the Biaka living range were defined for a 1970 census and genetic analysis of the Pygmies spending the winter in the southwest of the Central African Republic was performed. Pygmies censused were 3,410, and their numbers varied from 160 to 800 in each of the nine farmers' areas studied. There is some exchange with Pygmies living in adjacent regions of neighboring countries, Cameroon and the Democratic Republic of the Congo (Brazza), but not with Congo Kinshasa. The rivers Lobaye, Sangha, and Oubangui define the northern, western, and eastern boundaries of this group, but there are also other Pygmy groups north of the Lobaye who speak different dialects. The southern border of the censused area is not well defined, but it extends deeply into the forest of Congo Brazza. There is only modest linguistic variation within the Biaka, but there are exchanges with the most western Pygmy group, called Babenzelé.

The largest number of Biaka, circa 800, was censused (in 1968) in a farmers' village cluster called Bagandou, located in the Mbaiki province, south of the ferry across the river Lobaye. Bagandou is the name of the dominant Bantu farmers in the central part of this area, and the village cluster thus called is usually indicated on maps with the name of one of the major farmers' villages forming it, like Bokoma or Lombo. Farmers' villages of Bagandou had practically fused in 1970, forming a single, almost continuous cluster of huts, several kilometers in length around a central road in the shape of a T or Y. Many different languages are spoken in the same farmers' village, mostly, but not exclusively, Bantu, and they define fairly endogamous tribes spread over relatively wide areas much larger than any of the villages. The recent opening of new forest roads by western companies is changing the overall settlement pattern in this area.

Winter Pygmy camps are made of a much simpler type of huts, similar to those used in the forest, and are located either in the back of the farmers' huts, so that they are not visible from the road, or further away in the savanna or in the forest. In the nine areas there is an excess of the number of farmers versus that of Pygmies, with a ratio varying between 2:1 and 4:1. If we accept as deme the sum of all the Pygmy camps gravitating around a farmers' village or village cluster, then the average deme size calculated from village size, migration data, and formula 10.2 is 870 ± 108 (with a range from 564 to 1,616).

If the deme of Pygmies was considered the next higher population unit, that is, the tribe (identified linguistically), then the deme would be larger. But at the border of tribes camp clusters are often mixed, and the region occupied by a tribe is relatively large and difficult to examine thoroughly. The western part of the Central African Republic Biaka area exchanges with Pygmies calling themselves Babenzelé and speaking a different language or dialect. There are linguistic changes going on among Pygmies, and the boundary between dialects or languages is difficult to establish. At the northern periphery of the Biaka, the language used by Biaka was being replaced by that of the nearest Bantu farmers. In Biaka camps approximately midway between the two farmers' villages called Ndele and Bambio there was a clear split between the languages spoken by the old and the young moiety of the camp. Clearly, the transition to the new language, that of the local farmers, had happened 20–30 years earlier, and the old Pygmy language was understood only by the older moiety of the village. In Pygmy camps further north, Pygmy language had been completely replaced. The number of Pygmies per linguistic unit is difficult to estimate, and it is difficult to use as a deme the tribe defined linguistically. If the self-imposed tribe names are used to define demes, the deme size may be of the order of a few thousand.

The variance of gene frequencies of Pygmies in the nine areas could be computed and compared with that expected on the basis of the migration among areas. A good agreement of observed genetic variation and that expected under random genetic drift was observed (Zei and Zanardi 1986), giving some validity to the use of residential information for calculating deme size of the nine areas. There were no migration data or genetic data that would make it possible to repeat the computation using as deme size the larger linguistic units. There is also a problem in computing $F_{ST}$ from higher clustering levels, because the sampling error of variances is very high when they are calculated from a small number of observations.

## 10.4 ARE *COMUNI* (COMMUNES) DEMES?

It seems reasonable to conclude that, to some extent, the choice of a particular level of classification of human clusters as a deme is arbitrary. Nevertheless, in many cases there may be external criteria for using one level rather than another, and, in general, a useful level is that for which estimates of variances like $F_{ST}$ have the smallest relative error. This involves a choice of relatively small deme size.

Here we want to explore the advantages and shortcomings of using as demes the smallest administrative units in Italy, the *comuni* (plural of "comune," communes in English), for which there are usable "genetic" data in the form of surnames. There are approximately 8,400 *comuni* in Italy, with little variation over the years (those used in the 1993 list of Italian telephones were 8,216), and the average area occupied by each is comparable to a square of 6 km diameter (about 4 miles). The average population size is around 7,000, with enormous variation: circa 1,900 *comuni* had less than 1,000 inhabitants, 3,900 between 1,000 and 5,000, 1,800 between 5,000 and 20,000, while over 450 had over 20,000—but 53% of the population lives in these urban communities. Is the commune a reasonable choice for the geographic size of the deme? The choice of the commune may seem entirely arbitrary, but is somewhat attractive because it is the minimal population unit corresponding to a precise administrative center and for which statistical data of various sorts exist. It is also, to some extent, a social center.

As we have seen in the Parma example, in North Italy a commune is often made of several villages, and villages correspond to parishes unless they are very small. We have seen that there are 74 parishes and 12 communes (6.2 per commune) in the whole Parma Valley we examined. In North Italy one of the parish villages is often greater than all the others and is the administrative center of the commune because the office of the town mayor is located in it, as well as

TABLE 10.2. Number of communes per region, their average surface and population (1986)

| Region | Number of communes | Average surface (km²) | Average population | Density | Territory % M | H | P |
|---|---|---|---|---|---|---|---|
| Valle D'Aosta | 74 | 44 | 1,518 | 34 | 100 | 0 | 0 |
| Piedmont | 1,209 | 21 | 3,704 | 176 | 43 | 30 | 27 |
| Lombardy | 1,546 | 15 | 5,725 | 371 | 41 | 12 | 47 |
| Trentino | 339 | 40 | 2,576 | 64 | 100 | 0 | 0 |
| Venetia | 582 | 32 | 7,466 | 237 | 29 | 15 | 56 |
| Friuli | 219 | 36 | 5,635 | 157 | 43 | 19 | 38 |
| Liguria | 235 | 23 | 7,693 | 334 | 65 | 35 | 0 |
| Emilia-Romagna | 341 | 65 | 11,552 | 178 | 25 | 27 | 48 |
| Tuscany | 287 | 80 | 2,873 | 36 | 25 | 67 | 8 |
| Umbria-Marche | 246 | 39 | 5,741 | 146 | 29 | 71 | 0 |
| Latium | 370 | 46 | 13,518 | 291 | 26 | 54 | 20 |
| Abruzzi-Molise | 136 | 33 | 2,414 | 74 | 65 | 35 | 0 |
| Campania | 549 | 43 | 9,951 | 231 | 35 | 51 | 14 |
| Apulia | 257 | 75 | 15,064 | 200 | 2 | 45 | 53 |
| Basilicata | 131 | 76 | 4,657 | 61 | 47 | 45 | 8 |
| Calabria | 409 | 37 | 5,039 | 137 | 41 | 49 | 9 |
| Sardinia | 369 | 65 | 4,320 | 66 | 14 | 68 | 18 |
| Sicily | 390 | 66 | 12,581 | 191 | 24 | 62 | 14 |

(often) the school. In South Italy, communes are small towns, larger, on average, than in the north. There also are two administrative population units higher than the commune, the province and the region. Today there are 103 provinces, and thus there are 84 communes per province on average. There are 20 regions (or 18, since Marche and Umbria are often grouped into one, as well as Abruzzi and Molise, as in table 10.2). There are thus 5 provinces and 420 communes on average per region. The location of the provinces and the regions is shown in figure 10.1. The average surface and number of inhabitants of a commune and the distribution of mountain (M), hill (H), and plain (P) territory in each region are shown in table 10.2. The average surface of a commune varies from a minimum of 15 km² (Lombardy) to a maximum of 80 (Tuscany), and the number of inhabitants from 1,500 (Valle d'Aosta) to 15,000 (Apulia).

Might the commune represent a "random mating area," within which a spouse is mostly chosen? It would then be an appropriate

FIGURE 10.1. Regions and provinces of Italy.

choice for a deme. For all communes we have surname data and, in particular, their "Fisher's surname abundance" $N_S$ (the estimate of $Nm$ from surnames), to be used for the measurement of drift, as was already done in chapter 5, in ways to be further discussed in the next sections.

One must, however, consider that there are several confounding factors. The Italian communes have been subject, in the last hundred years or more, to important demographic changes, which are continuing. Little has been done to correct this situation, as has happened in other European countries, although a certain number of the smaller, disappearing *comuni* have been canceled, and new ones made by splitting older ones. In fact, a large number of *comuni* located in mountain or hilly areas have been in the last hundred years subject to progressive depopulation. This is largely true of those that are now under 1,000 inhabitants, and estimates of drift based on surname abundance will be potentially lower in them than they most probably were a century ago. Among the communes exposed to population erosion, especially in mountain areas, some became touristic centers, where many tourists have established a second house. Major touristic centers, located along the sea and lake coasts or in mountain resorts now have a large population during the tourist season and shrink considerably during the rest of the year. Estimates of deme size in such *comuni* will inevitably be artificially inflated.

At the upper end of the size distribution, a very few cities have increased enormously in size in the last century. They are all still a single commune. Most medium-size cities have increased only modestly in number of inhabitants, unless there was a major increase of industry or tourism. Cities can hardly be considered areas of random mating, especially as there is higher socioeconomic and residential segregation in them, compared with the smaller centers. But basically almost half of the Italian population lives in centers with less than 20,000 inhabitants.

## 10.5 THE NEGATIVE CORRELATION BETWEEN $N$ AND $M$

The effect on drift of the size of a community, $N$, is measured by the product $Nm$, where $m$ is for the proportion of immigrants per generation. An analysis of the dependence of $Nm$ on $N$, with $Nm$ measured by $N_S$ (Fisher's abundance of surnames discussed in chapter 5), is shown for the island of Sardinia (Zei et al. 1986, Lisa et al. 1996) in figure 10.2. Data used came from 40,000 individuals of all consanguineous marriages that took place in 1930–1959. As already stated before, only wives' surnames were used from each marriage, as marriages are usually celebrated in the wife's residence or birthplace. The data show the abundance of surnames as a function of $N$ after grouping the same observations by different criteria. The black circles on the left are the smallest population units employed, and correspond to the communes of Sardinia. Taken in isolation, they show very little, if any, correlation with $N$. In fact, for the black circles

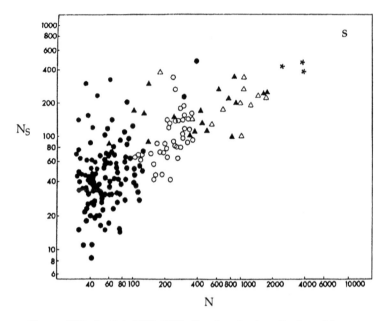

FIGURE 10.2. Sardinia 1930–1959. $N_s$ values (on logarithmic scale) computed for all communes having $N > 30$ (●), geographical neighborhoods (○), historical zones (▲), dioceses (△), provinces (*), and the whole island of Sardinia (S). All $N_s$ values plotted against the corresponding $N$ (sample size) scaled logarithmically. *From Zei et al. [1986].*

alone, the correlation coefficient $r$ between surname abundance and $N$ is 0.112, $P > 0.20$. Here $m$ varies from one commune to the other in inverse proportion to $N$, so that $Nm$ tends to remain practically constant and independent of $N$.

Something similar was noted in the upper Parma valley (chapter 5); moreover, in table 10.1, we saw that $F_{ST}$ varied almost in inverse proportion to $N$, as would be expected if $m$ were uninfluenced by $N$. Where the $N$ values of parishes were small, immigration was somewhat higher, but not enough to correct for very small values of $N$. In the case of the whole of Sardinia, it seems the negative correlation of $N$ and $m$ is sufficient to make $Nm$ approximately independent of $N$. There is still considerable variation of $Nm$ from commune to commune, and therefore drift varies considerably from one commune to the other, but the effect of the size of the commune seems to be largely offset by $m$.

Parishes and communes very largely overlap in Sardinia (there were 405 parishes and 349 communes at the time this work was carried out) and therefore the grouping by parish (which was not examined) and that by commune should give very similar results. We examined higher clustering levels, also shown in figure 10.2. In the religious organization the diocese is the next level above the parish, and there are 11 of them in Sardinia. There is a large gap between the largest commune or parish and the smallest diocese, and it was filled by creating an intermediate population unit, a geographic neighborhood, pooling 4–8 communes from the same diocese according to their geographic locations. The clustering unit above communes in the civil hierarchy is the province. The three provinces existing at the time were Cagliari (CA), Sassari (SS), and Nuoro (NU). We also added the value of $Nm$ for the 18 historical zones and for the whole island of Sardinia (S, at the extreme right in figure 10.2).

It is clear that $Nm$ as measured by surname abundance increases with the size $N$ of the population considered, but the slope of the increase is much less than it would be if $m$ were constant. Clearly, $m$ decreases as $N$ increases. When $m$ is calculated as $\nu$ in the Fisher theory given in Zei et al. (1983c), it decreases with increasing $N$ as

$$m = 24N^{-0.29} \qquad (10.3)$$

This shows that the choice of an area for calculating abundance of surnames, or using other methods of estimating drift, is not arbitrary. It will affect the measure of drift, and $Nm$, instead of $N$, should preferably be used in (10.1). At the moment, however, the commune remains the best choice for the smallest unit to be considered a deme. Parishes would give very similar results in Sardinia, but in other parts of Italy parishes tend to be three times smaller than communes, on average. The main limitation of parishes is that there are very few statistical data about them. This analysis refers to data from the first half of the twentieth century and earlier, before the beginning of the major touristic development of the island. As usual, islands tend to be slow on average in their socioeconomic and cultural evolution and Sardinia is no exception, although in some respects it is more advanced than all other Mediterranean islands. But data from those times are probably still representative of earlier marriage customs.

Lisa et al. (2001) carried this analysis in Sardinia further, using two sets of data: surnames from consanguineous marriages and those, more numerous, from 1978 thelephone directories. They have studied the regression of Fisher's surname abundance $N_S$ on population size $N$ (both on log scale) for each type of territorial subdivision, as described before (including the linguistic one), and found

$$N_S = 0.64N^{0.57} \qquad (10.4)$$

They found a very similar relationship between $F_{ST}$ calculated from Sardinian surname frequencies and $N$, except that, since $F_{ST}$ is an in inverse relationship with surname abundance (and, in general, with $Nm$), the relationship is now

$$F_{ST} = 2.58N^{-0.61} \qquad (10.5)$$

The exponent of $N$ in (10.4) and (10.5) is less than 1 and is approximately 0.6 but with an inversion of the exponent sign in the two equations. This indicates that $Nm$ is approximately proportional to the reciprocal of $F_{ST}$, as expected, because of (10.1). The regression of $Nm$ on $N$ for the communes is almost zero in figure 10.2, and therefore the immigration into a commune must decrease in Sardinia with the size of the commune as approximately $N^{-0.4}$

Lisa et al. (2001) have also extended similar calculations, using surnames from consanguinity data obtained from the Vatican archives (chapter 9) to all 200 continental Italian dioceses, and added data for 61 provinces in existence at the time the consanguinity data were collected (1911–1964). Results are presented in figure 10.3 and clearly show that three of the biggest Italian cities, Rome, Turin, and Milan, are out of line with the rest of the provinces. They must have exercised a great migratory pull already in the first half of the twentieth century. The overall relationship between surname abundance and $N$, ignoring the three major outliers, is quite similar to that observed in Sardinian communes in figure 10.2.

One cannot exclude that consanguineous marriages have a slight bias compared with general marriages, because consanguineous matings are especially frequent in rural and agricultural parts of the country, and there may be a custom of increased sedentariness dictated by attachment to the land. Consanguineous marriages also have special merit for keeping family land property together. We have

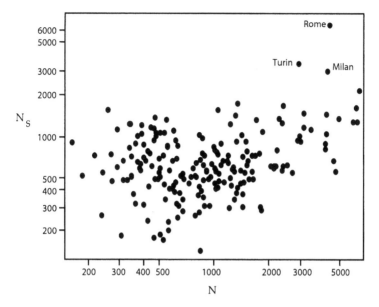

FIGURE 10.3. Italy 1911–1964. Relationship between $N_s$ and sample size $N$ in 200 dioceses.

therefore repeated the analysis on the general population, using *comuni* as demes, as reported in the next section.

## 10.6 USING SURNAMES FOR EVALUATING DRIFT

As an introduction we give in figure10.4 the frequency distribution of all Italian surnames obtained from phone books of year 1993, the source of our collection of Italian surnames, courtesy of SEAT. The total number of private telephones in Italy in 1993 was 18,554,690, and the total number of different surnames from the private telephone files was 332,525 (55.8 telephones per surname, on average). The number of telephones gives an approximate estimate of the number of Italian families. The approximation is generated by multiple houses belonging to the same families or their members; also, a small fraction of households cannot afford to have a telephone. Very few telephones are unlisted; it is more common that, for reasons of privacy, the telephone is listed under the name of another relative with a dif-

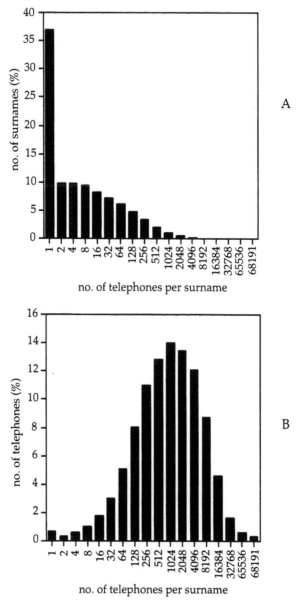

FIGURE 10.4. (A) Frequency distribution of Italian surnames from telephone books of 1993 per frequency class of surnames. (B) Frequency distribution of telephones (households) having surnames belonging to the same frequency class.

ferent surname or name. Thus, the list of telephones may lose a few members from the upper socioeconomic layers, and a few from the lower ones, but the coverage of the population is likely to be remarkably good.

Figure 10.4A shows the distribution of the number of surnames, from 1 to the maximum frequency of 68,191. Numbers of surnames are grouped in classes of powers of 2. The total numbers of households (telephones) having surnames belonging to the same frequency class are shown in figure 10.4B: the distribution is close to log-normal. Both distributions show that the very rare surnames form a very small mode of their own. Many of these are foreign and mostly of recent introduction in Italy. Singletons (names appearing only once) are clearly in excess, and were excluded in the analysis per commune.

Most surnames arose in Italy after the Council of Trento (1545–1562), during which the Roman Catholic Church imposed the keeping of parish books for records of baptisms, marriages, and deaths, and with them the generalization of the use of surnames. An unknown fraction, but most probably a minority, arose before the Council of Trento, between the eleventh and the fifteenth centuries. They can be found in the earliest books of baptisms of major churches or in notary public records. Usually these are disproportionately of noble or rich families. Other surnames are younger, as, for example, those indicating illegitimate birth (Deodato, Degli Esposti, or simply the name of a woman, most probably the mother) or those of recent foreign immigrants. In the last part of the sixteenth century the spelling of surnames changed somewhat—some were, for instance, given originally in Latin, then translated into Italian (e.g., De Ambrosiis, then Ambrosii or Ambrogi). But because demographic records were maintained by priests, who received a fairly rigorous education in seminary schools, including Latin, and both Italian and Latin languages are mostly spelt phonetically, spelling errors seem very rare.

Most surnames in Italy are patronymic. They started as the name of the father, in the genitive Latin form, typically ending with -i: Alberti is the surname originally given to the son of Alberto that became frozen when the father's surname was regularly transmitted to children. There are many varieties originating from one name when the father's name had changed to a diminutive or underwent other modifications (Albertini, Albertinetti, Albertelli, Albertacci, Albertoni). Each of these surnames may have had independent, multiple origins. Many patronymics are clearly derived from germanic first names (including Alberto), indicating the high regard for Goths, Longobards, and Carolingians, who conquered parts of Italy beginning with the sixth century A.D. But there are also names indicating a job or profession, a nickname, or a placename.

Common surnames are especially likely to have had multiple origins (polyphyletic). The most common surname, Rossi, is carried by 68,191 families (0.37% of all Italian families, or 1 in 272). It comes from "red hair," and is most common in North Italy, where there is the highest proportion of red-haired individuals. This is not surprising, given that red hair is especially common in areas where Celtic languages were (and still are today) spoken, and they were largely spoken in the north of Italy until the Roman conquest in the first centuries B.C. But the frequency of red hair is only slightly different in most Italian regions. We have records of its frequency in conscripts of the last years of the nineteenth century (Livi 1896), and the average frequency of red hair (0.57%), as well as the global one of surnames meaning red hair (Rossi, Rosso in northern Italy, Russo in southern Italy, Ruju, Ruggiu in Sardinia, and others) are similar and do not vary much from region to region, except for Sardinia where red hair and the relevant surnames are particularly rare (0.24%).

Some nicknames also are very frequent, but none is as frequent as those derived from red hair: Biondi (blond), Bruni (brown), Bianchi (white). Other somatic traits are less frequent: Curti (short), Gigante (giant). Some are derogatory: Malatesta (bad head), Mezzatesta (half-head), Lapelosa (hairy woman), Dentamaro (bitter tooth).

The most common job surname is Ferrari (39,164 families), with many variants all referring to *ferro* ( = iron), and other relevant roots (Fabbri, the equivalent of Smith) and other related ones in English, Fabre in France, and so on. Their frequency may reflect approximately that of the professionals who are so named.

Analysis of the geographic distribution of surnames indicating a place, *toponyms*, in Sardinia showed that the toponym tends to be at the geographic center of the distribution of the surname. At some time the place may have been named after a family living in it, who multiplied and spread outside. One might expect that a toponym arising from a rare place arose only once, but the distribution of the

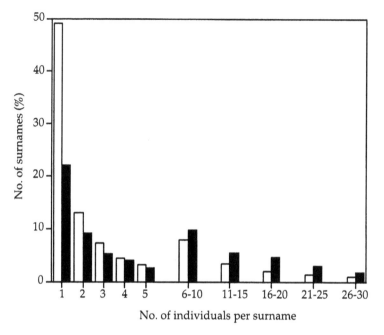

FIGURE 10.5. Sardinia 1984, data from electricity bills. Frequency distribution of all surnames: white columns. Frequency distribution of surnames recognized by a toponym: black columns.

numbers of carriers of toponyms shows little difference from that of general surnames (figure 10.5), except for a discrepancy for surnames represented only once. These data for Sardinia were obtained from electricity bills kindly provided by the ENEL company, and population coverage is more complete than for telephones. Toponyms seem to follow the expectations of a regular distribution of surnames, but there is a great excess of singletons among general surnames, and not for toponyms. They probably represent mostly recent immigrants (see also figure 10.4). The slight excess of toponyms in the right part of the distribution of figure10.5 must be simply due to the excess of singletons in the general distribution.

The extreme variation of number of telephones per surname is a strong indication that surnames arose repeatedly (are polyphyletic). Such surnames arose more frequently because there were more opportunities, and had a higher "mutation frequency" (e.g., Rossi, and related ones, Ferrari). The extent of polyphyletism may have an influence on the calculations of drift made from surnames, which may be neutralized by further research. On the other hand, variation of mutation rates is clearly present also in DNA nucleotide sites. It is possible that the mathematical theory suggested to describe the variation of genetic mutation rates can also be used for surnames. In the absence of adequate work on the relevant problems, we will use unchanged the basic measure of drift we have already employed in the upper Parma valley, surname abundance $N_S$.

The use of surnames for the study of drift was already justified in chapter 5 on the basis of data from the Parma valley. We have dedicated in the past much attention to the use of surnames in lieu of alleles at a locus for the measure of random genetic drift. In particular, in the work by Yasuda et al. (1974) we have shown the application of the Karlin and McGregor (1967) distribution of the numbers of alleles to that of surnames from parishes of the Parma valley. The goodness of fit of the distribution and the accuracy of its use to calculate $N$ and $m$ was remarkable. In the same paper we also showed the usefulness of the distribution of progeny size (a geometric with augmented proportion of zero children) in conjunction with the same distribution to estimate the probability of extinction of surnames.

In the article by Zei et al. (1986) we have studied the distribution of surnames in Sardinia and shown that one can use the Fisher distribution as an approximation to the Karlin and McGregor distribution, which is easier to use. We have also shown that, especially in samples from areas under economic development, and especially cities, the frequency of surnames represented in the sample by very few individuals (one or at most two) can be exagerated because of immigration of people of more distant origin, and given formulas for correction after elimination, especially of single surnames. The same anomaly was noted in an extensive analysis of many samples of the Chinese population (Du et al. 1992). For these reasons we have adopted throughout $N_S$, the Fisher abundance of surnames, for measurement of drift, and in the following we have tested for the effect of correction by elimination for single surnames in the analysis of Italian *comuni*, although it has a modest effect. Measures of drift for the Y chromosome, mtDNA, and other genes could follow similar routines, but this will require new research.

We have, here, a first chance to make a comparison between $F_{ST}$ for autosomal genes, surnames, and surname abundance in the Parma valley using results in tables 5.9 and 5.10. We expect $F_{ST}$ for genes calculated among parishes to be equal to $1/(1 + 4Nm)$, and that for surnames to equal $1/(1 + Nm)$. The $Nm$ value thus calculated from $F_{ST}$ variance of genes or of surnames should be the same as that obtained from $N_S$. We have two independent estimates for each of the three values: one from the plains + hills (P + H) and one from the mountains (M). The observed values are

|                              | P + H | M    |
|------------------------------|-------|------|
| $Nm$ from $F_{ST}$ for genes    | 32.6  | 11.1 |
| $Nm$ from $F_{ST}$ for surnames | 16.3  | 8.6  |
| $Nm$ from $N_S$                 | 28.5  | 11.9 |

There is a discrepancy between the three different $Nm$ values obtained in the plains and hills. Here drift is lower by a factor between 2 and 3 compared with the mountains, and therefore $Nm$ is higher than in the mountains. In fact, it is too low to be detected as genetic variation above that expected because of statistical noise, and the greater variation of $Nm$ calculated above for P + H indicates perhaps that there is a high error in estimating drift when it is very low. Estimates based on variances like $F_{ST}$ have particularly high standard errors, while it is possible that an estimate like $Nm$ from $N_S$ is statistically more robust than that obtained from $F_{ST}$-like variance. The $Nm$ value of P + H obtained from $N_S$ is 2.4 times higher than that for M, and this is in agreement with what we saw by other methods in chapter 5. It is quite encouraging that the estimates of drift in the mountains, where we have ascertained that drift exists and is detectable statistically, give nearly the same value with three different methods, two using surnames and one using gene frequencies.

We know that there exists an estimate of $Nm$ from genes based on the number of alleles, which is an equivalent of $N_S$, but it gives satisfactory results only when the number of alleles is high. Because the variety of surnames is always high, it is likely that it can be statistically robust. When more examples of estimation become available with both surnames and genes, and different types of genes, it may be possible to confirm that surnames, when they have been available for a sufficiently long time in the population, may provide a very efficient method of estimating genetic drift. Of course, estimates from surnames are valid only in the linguistic areas of origin, and only if their origin is sufficiently old. In some populations surnames arose very recently—for instance among Taiwan aborigenes (Chen and Cavalli-Sforza 1983) and in many Chinese ethnic minorities (Du et al. 1992). In such situations surnames provide little help. But in Japan, where surnames arose fairly recently (with the Meiji restoration in 1868) they have been used satisfactorily for some purposes.

## 10.7 A DRIFT MAP OF ITALY BY COMMUNES

We show in figure 10.6 a geographic map of Italian communes. For each of them we give the surname drift estimate $N_S$ of 8,092 communes calculated from the 1993 list of telephones. Dark boundaries separate the 20 regions and thin white boundaries the 103 provinces. Two-letter acronyms mark the cities that are the head of each region. There are six classes of Fisher's surname abundance ($N_S$), measuring drift, distinguished by tones of gray. We have seen that the distribution of surname abundance is very closely log-normal (figure 10.4), and therefore the six classes are on a logarithmic scale. The logarithmic values of the class boundaries and the central class values are given in table 10.3 for the distributions of communes in the twenty regions. As mentioned before, singleton surnames were excluded, for reasons discussed in more detail in other papers (Du et al. 1992, Zei et al. 1986, Cavalli-Sforza 2001), but the effect is hardly visible.

The commune having leading attributes in the administration of each province is a city of some numerical and economic importance and its two-letter symbol is indicated inside each province in figure 10.1. It always has a large $Nm$ value, but does not necessarily appear in the first, and sometimes not even in the second, highest classes. All communes that have a high $Nm$ value, although devoid of administrative importance and peripheral in a province territory, are towns that recently developed financial importance, mostly for touristic but also for industrial or commercial reasons. Table 10.3 shows 110 communes (1.4%) in the densest class. They are mostly cities that are heads of provinces, and together with the second highest class (4.8% of all communes) are practically all the cities of importance. The two lowest classes (respectively, 13.4 and 30.2% of the communes) include practically all the genetic isolates of Italy—although the territory of some true isolates may not coincide exactly with that of the communes where they are located. The communes in the two smallest $Nm$ (or greatest drift) have a geographic distribution that tends to point to the crest of the major mountain chains, the Alps and the Apennines (figure 10.6), since genetic isolation tends to be highest in the mountains and to some extent even in some hilly regions, but some other communes have small $Nm$ even if they are not in the mountains. The Alps form an almost complete circle isolating North Italy from the rest of central Europe. The triangle carved in the north of the country by the boundary with Switzerland, whose southern tip is not far from Milan, is the Italian-speaking Swiss canton Ticino, located south of the Alps, data for which are not available in our map survey. The southwestern end of the Alps continues in the chain of the Apennines that first encircles Liguria, a very montainous region, near the coast, and then continues south, forming the backbone of the whole Italian peninsula and continuing into Sicily. We have seen in chapters 4 through 7 the strong effect of the low density of populations as a function of altitude on both consanguinity and drift. The relative importance of mountains in each Italian region is shown in table 10.1.

The major center in Piedmont, the first region in the northwest of Italy, is Turin, the capital of the region. With Milan and Genoa, the other most important industrial centers in Italy, it forms the so-called industrial triangle. Other cities are of considerable agricultural and industrial importance. The high $Nm$ value at the extreme west of Piedmont is a major skiing area.

Northwest of Piedmont is the Valle d'Aosta, closed in the mountains and with a strong French cultural background and major skiing and mountaineering areas. Genetic isolates in the west and north follow the western part of the Alps circle.

Next northeast is Lombardy, where Milan is the principal commercial center of Italy and is rich in industry. It and Piedmont have the greatest concentration of economically developed communes, which

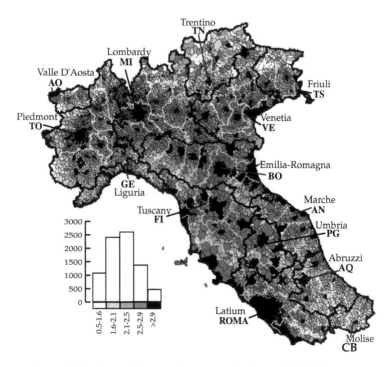

FIGURE 10.6. Geographic map of surname abundance $N_s$ in Italian communes obtained from telephone books in 1993.

makes them most dense genetically. They are followed in the scale of wealth by Tuscany, Emilia, Latium, Liguria, and the Venetian regions. The northern cities of Lombardy are more industrial, and the southern ones more agricultural.

Of the three Venetian regions, Trentino is entirely mountainous, with important touristic and economic development, and has a strong German-speaking component of Southwest Austrian (Tyrolean) origin in its northern part. Venetia is mostly in the plains and together with Friuli-Venezia Giulia has had a major industrial development since World War II. Genetic isolation in the eastern part of the Alps is not as conspicuous as in the western part, also because of the extraordinary touristic development.

Liguria is a narrow region, almost entirely mountainous, along the coast of the northern Tyrrhenian Sea. It includes the first part of the Apennine chain, south of Piedmont and northern Emilia. It is dominated by Genoa, the leading port and industrial city at the center of the region. La Spezia and Savona are ports second in importance to Genoa, and Imperia and San Remo are touristically important, like practically all the Genoese Riviera. Major isolates are in the internal mountains, but all the coastline is practically a sequence of smaller and bigger sea resorts, wherever there are good beaches.

Emilia is a long region extending from Lombardy to the Adriatic Sea. Most of its provinces are in the very fertile plains extending from the Po River, forming the northern boundary, to the crest of the Apennines, forming the southern boundary. The southern mountainous area is relatively barren, but in between the mountains and the plains are gentle hills. Cities are, from west to east, Piacenza, Fidenza, Parma, Reggio Emilia, Modena, Bologna, Ravenna, and Rimini; north of Ravenna is Ferrara. Bologna is the leading city and

FIGURE 10.6 continued.

TABLE 10.3. Distribution of the communes per $N_s$ values in the twenty Italian regions

| Region | $N_s$: 12 / Log $N_s$: 0.5–1.65 | 75 / 1.65–2.10 | 210 / 2.10–2.55 | 595 / 2.55–3.00 | 1680 / 3.00–3.45 | 8400 / 3.45–4.40 | Total | Average ± SD |
|---|---|---|---|---|---|---|---|---|
| Valle D'Aosta | 15 (20.3) | 25 (33.8) | 16 (21.6) | 16 (21.6) | 1 (1.4) | 1 (1.4) | 74 | 2.087 0.5413 |
| Piedmont | 177 (14.7) | 429 (35.5) | 372 (30.8) | 161 (13.3) | 54 (4.5) | 15 (1.2) | 1208 | 2.132 0.4996 |
| Lombardy | 137 (8.9) | 342 (22.1) | 577 (37.4) | 374 (24.2) | 82 (5.3) | 33 (2.1) | 1545 | 2.306 0.4946 |
| Trentino | 119 (35.6) | 120 (35.9) | 74 (22.2) | 14 (4.2) | 4 (1.2) | 3 (0.9) | 334 | 1.825 0.4966 |
| Venetia | 22 (3.8) | 105 (18.0) | 289 (49.7) | 134 (23.0) | 27 (4.6) | 5 (0.9) | 582 | 2.344 0.3899 |
| Friuli | 29 (13.2) | 56 (25.6) | 78 (35.6) | 42 (19.2) | 9 (4.1) | 5 (2.3) | 219 | 2.218 0.5118 |
| Liguria | 31 (13.2) | 57 (24.3) | 56 (23.8) | 50 (21.3) | 29 (12.3) | 12 (5.1) | 235 | 2.356 0.6197 |
| Emilia-Romagna | 2 (0.6) | 30 (8.8) | 129 (37.8) | 146 (42.8) | 26 (7.6) | 8 (2.4) | 341 | 2.547 0.3456 |
| Tuscany | 4 (1.4) | 31 (10.8) | 90 (31.4) | 119 (41.5) | 36 (12.5) | 7 (2.4) | 287 | 2.590 0.4042 |
| Umbria | 4 (4.4) | 16 (17.4) | 46 (50.0) | 20 (21.7) | 4 (4.4) | 2 (2.2) | 92 | 2.364 0.3928 |
| Marche | 10 (4.1) | 77 (31.3) | 108 (43.9) | 37 (15.0) | 13 (5.3) | 1 (0.4) | 246 | 2.255 0.3741 |
| Latium | 5 (1.3) | 126 (33.6) | 151 (40.3) | 61 (16.3) | 30 (8.0) | 2 (0.5) | 375 | 1.864 0.5715 |
| Abruzzi | 11 (3.6) | 132 (43.3) | 125 (41.0) | 30 (9.8) | 7 (2.3) | 0 (0.0) | 305 | 1.683 0.4696 |
| Molise | 9 (6.6) | 89 (65.4) | 31 (22.8) | 5 (3.7) | 2 (1.5) | 0 (0.0) | 136 | 1.499 0.3855 |
| Campania | 90 (16.4) | 214 (38.9) | 141 (25.6) | 78 (14.2) | 26 (4.7) | 1 (0.2) | 550 | 2.106 0.4640 |
| Apulia | 6 (2.3) | 71 (27.6) | 113 (44.0) | 60 (23.4) | 5 (2.0) | 2 (0.8) | 257 | 2.287 0.3807 |
| Basilicata | 16 (12.2) | 69 (52.7) | 37 (28.2) | 7 (5.3) | 2 (1.5) | 0 (0.0) | 131 | 1.999 0.3170 |
| Calabria | 66 (16.1) | 211 (51.6) | 103 (25.2) | 22 (5.4) | 7 (1.7) | 0 (0.0) | 409 | 1.985 0.3542 |
| Sicily | 15 (3.9) | 133 (34.1) | 150 (38.5) | 79 (20.3) | 10 (2.6) | 3 (0.8) | 390 | 2.248 0.3859 |
| Sardinia | 149 (39.6) | 155 (41.2) | 48 (12.8) | 17 (4.5) | 7 (1.9) | 0 (0.0) | 376 | 1.803 0.4137 |
| Total numbers | 1,082 (13.4) | 2,444 (30.2) | 2,612 (32.3) | 1,457 (18.0) | 387 (4.8) | 110 (1.4) | 8,092 | |

*Note.* Percentages are in parentheses. SD, standard deviation.

the seat of the oldest, still very active university, but all of Emilia has taken considerable economic initiatives. The communes in the Apennines have relatively low population densities, but only a few of them have a real poverty of surnames. The Adriatic coast of Emilia is an extremely active and dense sea resort on a 50-km-long coast, with 200,000 fixed residents and 2 million tourists, mostly in the summer.

Tuscany borders Emilia in the north and Umbria and Latium in the east, and has a long coastline on the Tyrrhenian Sea, which also has an almost uninterrupted series of sea resorts. As in Emilia, genetic isolation in the Apennines is barely visible. The mountains here have been a considerable attraction over the last millennia because of metals and marble quarries, and so has the island of Elba, which is rich in iron. Etruria, in south Tuscany and northern Latium, became a powerful and active center of economy, engineering, and culture in the first millennium B.C. and strongly influenced the initial development of Rome, later falling under Roman control. In the Renaissance the city of Florence and the rest of Tuscany were the major center of European finance, industry, art, and science (Galileo), but in the early seventeenth century, local economy began to slow down (Cipolla 1952). Lucca, Pistoia, Pisa, Arezzo, and Siena were also active art and literature centers. Livorno was a major port of the Mediterranean and the entry to southern Europe for Jews after they were banned from Spain in 1492.

Umbria shared with Tuscany a strong development in the Renaissance, and so did, though to a lesser extent, Molise and Abruzzi. These regions are hilly or mountainous, with good agriculture and food and excellent tourism.

Latium was the dominant region in Italy, beginning with the development of Rome in the eighth century B.C. and continuing with the extension of its dominion to the Italian peninsula and part of the western and southern Mediterranean in the second half of the first millennium B.C., and to all the Mediterranean, all of western and southeastern Europe, reaching its peak in the first half of the first millennium A.D. The western Roman Empire fell to barbarian invaders from East Europe in the fifth century A.D. The modern development of Rome and Latium is mostly touristic, but south of Rome is an industrialized area.

Campania is dominated by Naples with its strong touristic and industrial development, but also has agriculture. Beginning here and extending to all the south, including Sicily, Greek colonies flourished beginning in the first millennium B.C., and Greek language is still spoken in small enclaves.

South Italy was called Magna Graecia, because it was demographically and even economically more advanced than the Greek homeland, as tourists who visit Greek ruins of Paestum, south of Naples, and many parts of Sicily well know. After the fall of the Roman empire Arab conquerors, beginning in the eighth century A.D., restored strong agriculture in Sicily, but the later domination of Normans and of Spaniards was highly exploitative, with the exception of a happy period in the thirteenth century when German emperors extended their domination to the south of Italy. Basilicata and Calabria and the northern part of Apulia are hilly or mountainous; Apulia is flatter and richer. Sicily has many major cities with some industrial development, but tourism is the most important economic source throughout most of the south.

Sardinia was settled in Paleolithic times, probably from the Iberian peninsula, and in the first two millennia B.C. a unique, strong Bronze Age culture built a large number of fortified towers (*nuraghi*). About 300 years ago the population was already made of about 300,000 individuals, almost one-fifth as many as today. Original settlers are still, as usual, more numerous in the most remote and mountainous part, in south-central Sardinia, where density is lowest. The strong density in the northwestern area is part industrial, part touristic. Other parts of the coast are well known from the touristic point of view.

## 10.8 STATISTICAL OBSERVATIONS ON THE ITALIAN DRIFT MAP

The variable on which the drift map of Italy is based, $N_S$ values, the equivalent of $Nm$ calculated from surnames on the basis of surname abundance, has been transformed into its decimal logarithm because this transformation has a nearly perfect normal distribution. There is only a slight excess of values at both ends. At the lower end, the few foreign surnames are probably the major cause. At the upper end of $Nm$ values are cities that showed exceptional growth, some of which are clearly visible in figure 10.6.

TABLE 10.4. Relationship between surname drift ($N_S = N_m$ calculated from surnames) and various statistics

| Correlation between | Correlation coefficient | Regression coefficient ± SD | |
|---|---|---|---|
| $N_S$ estimate and $m$ | − 0.398 | − 0.194 | 0.0050 |
| $N_S$ and 1991 population census, $N$ | 0.807 | 0.724 | 0.0069 |
| $N_S$ and population density | 0.659 | 0.595 | 0.0755 |
| $N_S$ and altitude | − 0.439 | − 0.384 | 0.0087 |
| $N_S$ and agriculture | − 0.402 | − 0.320 | 0.0102 |
| $N_S$ and industry | 0.318 | 0.421 | 0.0140 |

Note. SD, standard deviation.

Almost all the correlations of $N_S$ with standard statistical variables, using the logarithmic transforms are remarkably normal, with very few outliers at both ends of the distribution. There are only minor deviations from nearly perfect linear regressions.

The most interesting results are summarized in table 10.4, where we report the correlation coefficient, the regression coefficient, and its standard error of $N_S$ values with several statistical variables available for the communes. The statistical significance of the correlation coefficient when compared with zero is always very high ($P <<$ 0.0001). This is not surprising given the high number of data (over 8000) and the good linearity of the regressions.

There is a high correlation between our estimates of $Nm$ from surnames in all communes and the corresponding population census values taken in 1991. The relation between their logarithms is fairly linear but tends to flatten at the two extremes. The correlation is much higher than the low ones calculated in Sardinia in (10.4) and (10.5) between $N_S$ and the size of the communes, obtained on surnames derived from consanguineous matings. It is possible the matings used in the first survey are constrained to a narrower range, because consanguineous marriages are not usually found in a wide area, but it is more likely that the difference between the correlations of $N_S$ and population size in figures shown in section 10.4 and in table 10.4 reflects the difference in time, since data in table 10.4 are from more than half a century later than those of (10.4) and (10.5). The regression of migration ($m$) on $N_S$ is here − 0.194, definitely smaller than that estimated from consanguineous mates in section 10.5 (− 0.4). We also find a negative correlation of $N_S$ with agricul-

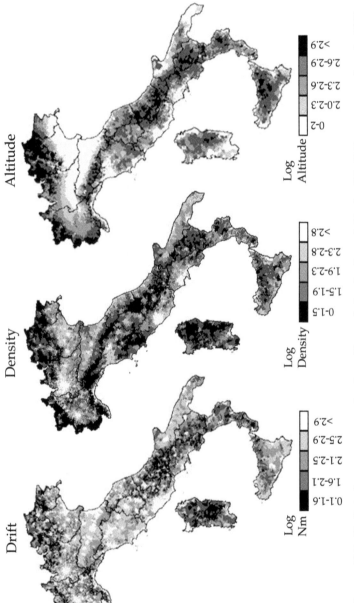

FIGURE. 10.7. Similarity between surname abundance $N_s$, population density, and altitude of the Italian communes from the 1991 census.

ture ($r = -0.402$) and a positive correlation with industry ($r = 0.318$), and, even greater, a negative one with altitude ($r = -0.439$) and a positive one with population density ($r = 0.659$). The simplest interpretation is that the older environment at the beginning of the twentieth century was richer in agriculture and poorer in industry, there was less migratory movement of people, and consanguineous couples, which are usually located at a lesser distance, were more likely to marry. In the second half of the century there was continuous economic and social change, and the correlations of table 10.4 reflect the changing times and the higher geographic and social mobility thus generated.

Finally, figure 10.7 shows the geographic map for the Italian communes of the three major factors, *drift*, *population density*, and *altitude*, designed in nonconventional scale to emphasize the similarity between them. Observed drift is represented with color intensity varying inversely with log *Nm;* color intensity of population density varies inversely with log density, and altitude color intensity varies directly with log altitude.

CHAPTER 11

# Conclusions

## 11.1 HUMAN CONSANGUINITY

Consanguineous marriages vary enormously in human populations, under the influence of differences in customs and laws. Unions with the closest relatives (parent–child and sib–sib) are considered incest and punished in almost every culture if they occur, with few historical exceptions in some ancient dynasties in Egypt and Persia. But marriage with less close relatives, like uncle–niece, is accepted and even encouraged in a few cultures. The Judeo-Christian tradition strongly condemned incest and oscillated between avoidance and tolerance of less close consanguineous unions. First-cousin marriages are extremely common, ranging between 20 and 50% of all marriages in Moslem cultures over a very large area, from North Africa to India and Central Asia. Such customs probably antedated the spread of Arabs to North Africa and Spain beginning in the seventh century A.D., and they are condemned in the Koran. A geographic map of consanguineous marriages among Caucasoids (Figure 11.1) (Bittles 2001) shows a strong contrast between the low values of consanguinity in Europe and people of European origin around the world, and the high values in North Africa and much of west, central, and south India. In other parts of the world there is mostly a tendency to avoid close cousin marriages (with the exception of traditional Japan).

There is some natural tendency to avoid incest also in animals, most probably because of customs of differential sex dispersal, probably supported by natural selection. In many cultures there is awareness that marriage of close cousins increases the chance of serious diseases in the progeny. In the Decretum by Gratianus (circa 1150), a

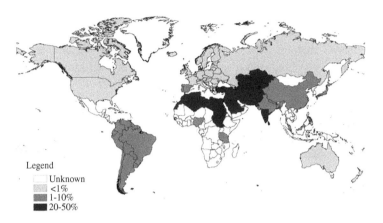

FIGURE 11.1. The current global prevalence of consanguineous marriage. *From Bittles [2001].*

monk from a monastery near Bologna, there is a specific statement on the high probability of deafmutism, blindness, and other severe handicaps in progeny from incest. In the United States there were pioneers like Bemiss (1858) and Arner (1908) who carried out relevant surveys of consanguineous matings to measure as exactly as possible the degree of damage. In the mid-nineteenth century Mantegazza (1868) published a survey of damage in the progeny of consanguineous parents. Many of his conclusions and hypotheses are correct to this day. He wrote: "Consanguineous marriages are more often harmful than harmless" and "the closer the family ties the more dangerous is the marriage to the offspring."

An analysis by Bittles and Neel (1994) has summarized the modern evidence, concluding that the average increase in prereproductive death rate in the progeny of first cousins due to genetic causes is 4.4. ± 4.5%. Unfortunately, the standard error of this estimate is very high because it is obtained by extrapolation, although it is based on a fairly large number of observations generated by many different authors in 38 populations from many parts of the world. It is very difficult, therefore, to say if the proportion of genetic deaths in first-cousin progeny varies between different populations. The estimate does not include early abortions, which are too difficult to evaluate reliably. From this figure one can infer approximately the damage expected in consanguineous matings of other degrees. It is expected to be almost twice as high in the highest degree of practiced consanguineous matings (uncle–niece and aunt–nephew, or double first cousins), four times lower in second cousin, and practically undetectable in third cousin matings.

In the great majority of countries in which consanguineous marriage is very frequent, hygienic conditions are still inadequate and therefore the total prereproductive death rate is usually large, in the worst conditions more than ten times higher than that expected for genetic reasons in first-cousin progeny. The effect of consanguinity can thus easily go unnoticed without careful study in these populations. Birth rates are very high in most of these countries and avoidance of consanguinity does not seem to have a high social priority. Even in developed populations the current consensus is that legislation prohibiting and/or punishing consanguineous marriage (still extant in some countries and in some American states) is not justified (Gibbons 1993). It is likely that the mutational genetic load accumulated over earlier generations may be decreased, and therefore the loss of progeny from consanguineous matings less heavy wherever consanguineous marriage has been practiced very frequently for a long time. The existing data collected and discussed by Bittles and his collaborators cannot exclude that genetic load varies in different populations. It is worth testing whether populations that have had a high frequency of consanguineous marriages for a long time have a decreased genetic load.

There remains the question of the causes of such remarkable differences in customs, leading to a remarkable variation in frequency of consanguineous marriages in various parts of the world. In particular, for first cousins the frequency range is greater than 100-fold. In some cases the forces at work are clear: particularly, low consanguinity may be due to effective avoidance, but high consanguinity may be due to a number of different causes, which are not always easy to dissect. Marriage with close relatives may be especially attractive where family values are especially important, the size of extended families is large, and social contacts are much more frequent with close relatives. In many countries marriages have long been arranged by parents, other relatives, or professionals, often at very young ages of the future bride and groom, and it is still standard custom, especially in developing countries, to arrange mariages rather than leaving them to the initiative of would-be spouses. In section 4.3 we hypothesized that the high frequency of consanguineous marriages in which the nearest ancestors (the parents) of the spouses are females is due to a tendency of women relatives to maintain family ties. This tendency also contributes to increasing the frequency of consanguineous marriages, though probably to a lesser degree than other factors. Contact with relatives is especially frequent in closely knit families, and in certain social environments there are also financial stimuli to keep families close. The extreme fractionation of the land owned by a single small owner, which followed the abolition of primogeniture, is often still observed today and suggests that it must have caused great concern and have led to some interest in consanguineous marriage, especially between first cousins. But this was certainly only a partial cause of increased inbreeding. There were other forces at work, and some of those connected with demographic transition may have been particularly powerful. One wonders if a tendency to positive assortative mating for social, economic, or other factors may also contribute to this phenomenon.

Our data indicate that there was a slow increase of consanguinity in Italy beginning with the first part of the nineteenth century. It was followed by a higher rate of increase toward the end of the century, until a peak was reached in most parts of Italy in the period between 1918 and 1925. The frequency peak was highest in the south and less pronounced in the north. A fast descent began after the peak, with downslopes greater in the north.

The phase of increase may have been due in part to the desire to keep family land together, since there is some coincidence between the beginning of the slow increase of consanguinity with laws against primogeniture promulgated in France by Napoleon and rapidly spreading to most of Europe. But the later and stronger increase is more probably due in greater part to the increased rate of population increment (the demographic transition), which manifested itself, in Italy, with very different patterns in different parts of the country. One can recognize four major types of demographic transition, marked by later and higher rates in the south and islands compared with the north and center. The increase of population increment caused an increase in the proportion of cousins in the population, and hence in the proportion of consanguineous marriages, with a delay of one generation for first cousins. The delay was expected to be longer for less closely related partners, but the phenomenon did not last long enough to manifest itself as clearly for them as it did for first cousins. In fact, at a later time other social forces, essentially increased migratory movements, caused a fall of consanguineous marriages, inaugurating a trend in the opposite direction. The regional variation in the height of the peaks in a selected set of Italian studies is in agreement with the indications of the regional variation of the patterns of the demographic transition. It is also possible that the political differences between the Vatican and the new state of Italy in the period 1870–1928 decreased the influence of the Catholic Church on marriage policies, but the data available on dispensations for consanguineous marriages seem to indicate that religious consent was almost regularly granted and carried at most some delay in marriage.

Our statistical knowledge of consanguinities for the whole country is complete only for the period from 1911 to 1964. It is therefore more informative for the period of the peak and the second phase, the descent of consanguinity. In this period we note very low values of consanguineous marriages with an almost flat trend in time in the Venetian region and in central Italy, a decrease from a relatively high value in the industrial triangle (Milan, Turin, Genoa), a slower decrease in the south, and high values with little tendency to decrease in the extreme south (Sicily and Calabria). The descent phase is clearly a manifestation of the same phenomenon described in Sweden by Dahlberg under the name of "breakdown of isolates": the increase in mobility, that is, of communication, transportation, and relocation of workers to newly arisen industrial centers. In Italy it took place about a half-century or more later than in Sweden.

It is particularly difficult to analyze and explain in detail a rapidly changing phenomenon in an extremely heterogeneous country that during the process went from a highly fragmented quasi-feudal condition lasting until 1860 for the north and south, and until 1870 for the center, to a modern, unified state. Nevertheless, it was possible to analyze in more detail certain parts and periods. We were helped by the availability of accessible demographic data: parish books for the last three centuries or more, and data of the Istituto Centrale di Statistica after 1865.

The importance of demography in this analysis is essential for evaluating the prior probability of marriage for six degrees of consanguinity: three "even" cousins (first, second, and third cousins) with equal length of branches, and three "uneven" consanguineous marriages with unequal length of branches (uncle–niece [UN] or aunt–nephew [AN], and 1½ and 2½ cousins). These six degrees can be further subdivided into a large number of pedigree types, defined by the sex of ancestors intermediate between the common ones and the consanguineous mates. The number of types of pedigrees varies from 4 to 64 for the six degrees. Each consanguinity degree and type of pedigree was found to have its own specific frequency of occurrence. We showed that it was possible to predict these frequencies on the basis of knowledge of two major factors: the age at marriage, which differs in the two sexes and affects the age at birth of children, and the probability of migration from the place of birth, which also differs with gender. Males migrate less than females in the northern region; they show a lesser difference in Sicily, but a variable one in different parts of Sardinia, depending on customs, occupation, and residence (pastoral or agricultural). The effect of age difference between spouses is more marked for uneven cousins, and that of migration for even cousins, but for an accurate prediction both factors must be considered jointly.

In addition to the age and migration effects, we found a social factor that can almost double the frequency of certain types of pedigrees: those in which the parents of the consanguineous mates in the pedigree are both women. This was observed regularly for all degrees of consanguinity. It indicates that mothers of spouses are important in arranging marriages and are part of the trend favoring consanguineous ones.

The knowledge of these factors allowed us to test if one can assume that in the area best investigated (northern Emilia) the choice of consanguineous mates can be considered to take place randomly, that is, with no preference for or against consanguinity itself. Computations indicated that there may be some avoidance, and an avoidance factor was calculated, which was especially important for the closest ties. But the conclusion on the existence of avoidance for closest ties depended very much on the assumption of complete reporting of consanguinities. When the area with highest consanguinity of the Parma valley was subjected to a more thorough analysis of expected consanguinities by a population simulation, there were indications of underreporting of the more remote consanguinities. This source of error might have inflated our estimate of avoidance of the closest consanguineous marriages. It seemed reasonable to conclude that in the northern area, which was most thoroughly investigated, the frequency of closest consanguineous marriages was relatively uninfluenced by fear of the negative consequences of consanguinity for progeny. In the south, even though the specific risk may have been known, it, and the need of requesting a dispensation for consanguineous marriage, were not deterrents strong enough to completely balance other, favorable factors. Possible real or perceived advantages of marriages within the extended family seemed to compete egregiously with the classical risk factors of consanguineous marrriages.

On the whole, our conclusion is that demographic factors like age preferences at marriage and migration allow a reasonable prediction of the frequency of consanguineous marriages, and that in Italy, especially today, avoidance, if present, is not major. There are, in addition, various social factors that affect negatively or positively the occurrence of consanguineous unions. Ecological factors like altitude, affecting village size, population density, and migration influence profoundly the overall frequency of consanguineous marriages. When data can be trusted to be sufficiently complete, and in Roman Catholic records they usually are, second-cousin frequencies are better indicators of the influence of demographic factors than first-cousin frequencies. First cousins are more sensitive to local preference or avoidance, but only exceptionally by a big factor.

The increase of all consanguineous marriages in the nineteenth century, slow at first and then more rapid until about 1920, is explained semiquantitatively by population growth during the demographic transition, but other factors, including the desire to avoid land fragmentation imposed by the elimination of primogeniture, are likely to have played a role. The rapid decrease of consanguineous marriages in the twentieth century is a consequence of the increase of internal migration, due to changes in communication, transportation, and jobs availability. But one must also acknowledge the importance of a general cultural change, beginning in the first half of the twentieth century, which might be described as increased general awareness, due to improved education, social contacts (e.g., because obligatory conscription helped decreasing illiteracy and favored exchange between north and south, thanks to the custom of sending conscripts from the north to southern regions and vice versa), and information (newspapers and radio, in the period considered) and increased working conditions and opportunities.

## 11.2 INBREEDING

The frequency of consanguineous marriages provides an estimate of average inbreeding of the population that is undoubtedly too low, by a factor perhaps as high as 2–5. The reason is the underestimation or total absence of knowledge on the more remote relationships. In countries where the rate of first-cousin marriages is very high this is not necessarily true, since first-cousin marriages can be ten to one hundred times more frequent there than, say, in European cities. A complication is the recent change in the rate of population increase, which affects inbreeding considerably. There are few empirical or theoretical investigations. Crow and Mange (1965) in research on a religious sect in rapid growth showed that inclusion of third and fourth cousins almost doubled the average inbreeding coefficients. Karlin and McGregor (1967) estimated theoretically the effect of population growth on the inbreeding coefficient. A.W.F. Edwards (1968) introduced and tested a method of estimating the average inbreeding coefficient by the Monte Carlo method; in the words of the author: "The simulation of gene flow through a given genealogy can provide the necessary information [about drift, etc.] as well as enabling inbreeding and kinship coefficient to be easily calculated using the concept of gene identity by descent."

A high inbreeding, if prolonged for a sufficiently high number of generations may decrease the genetic load of deleterious mutations (Cavalli-Sforza and Bodmer 1971, 1999). A possible example is Japan, where arranged cousin marriages were high until very recently. We do not know for how long the Japanese population has been inbred at a high level, but if this has been going on for a long time, it might have lowered the genetic load. This is indicated by the number of lethal equivalents found there (0.8), which seems lower than the current rough estimate of the average in the western world (1.4).

The relative frequency of recessive diseases is higher in inbred individuals and it is relatively higher the rarer is the recessive gene. The frequency of deleterious genes is determined by the balance of mutation and natural selection: it is higher the higher the mutation rate and lower the more deleterious is the gene. For recessive genes it also depends on the Darwinian fitness of the heterozygote. Very rare recessive diseases are detectable only in the progeny of consanguineous marriages. About half of the circa 10,000 Mendelian diseases known are recessive, and they are, with few exceptions, rare (with prevalence less than 1/50,000) or extremely rare, and have therefore been observed in only a few families.

The analysis of imbreeding effects on diseases by linking hospital records with our Italian list of circa 540,000 consanguineous marriages that took place in 1911–1964 have shown no unexpected results. They have improved existing estimates of the incidence in the country of relatively common recessive diseases. The estimates of mortality of inbred children, in particular for the progeny of first cousins, are a little lower (3.5 %) than the average reported by Bittles and Neel (4.4%), but the difference is not significant.

Conscription records always gave the names of the conscripts' father and mother. Linking conscription records collected in 1890 and 1891 with records of consanguineous marriages of the Diocese of Parma allowed us to study the effects of consanguinity on some quantitative traits measured at conscription: stature and chest girth. We found inbreeding depression for both. That for chest girth was rather clear-cut, while that for stature was obscured by an unexpected and surprising result: the progeny of third cousins was significantly taller than controls and all other inbred groups. Stature is known to be sensitive to socioeconomic class effects, and the inbreeding of third-cousin progeny is so low that the observation can hardly be a consequence of it, but is indicative of an unusual social origin of third cousins. Third cousins must have had stronger economic or cultural reasons for remaining in the same area, or at least for having kept closer family contacts and for having chosen to marry together.

The estimate of average inbreeding in Italian provinces, based on 1911–1964 marriages, is $1.31 \pm 0.39 \times 10^{-3}$. Because this is an underestimate by a factor between 2 and 5, the true value is likely to be between 2.5 and $6.5 \times 10^{-3}$.

The highest average inbreeding values, just above 2% around

1920, were observed in the Eolian Islands of Alicudi and Panarea. These had a number of inhabitants varying in the last century and a half between 112 and 813 for Alicudi and 236–790 for Panarea. The highest reported value in the world is 4.3 % for Samaritans.

## 11.3 GENETIC DRIFT

The Parma data were most probably the first instance in which observed values of random genetic drift, estimated on the basis of blood group genes, were compared with those expected on the basis of demographic factors determining drift (Cavalli-Sforza 1958). The area of the Parma valley was analyzed for the major demographic factors affecting drift: the size of social groups and migration. In this area, villages almost always correspond to parishes, demographic records for which were available for the last three and a half centuries before 1950. There was little variation of population size over this period; a slight decrease of size in the eighteenth century was compensated later in the century, and excess population generated in the demographic transition at the end of the nineteenth century was mostly absorbed in emigration abroad or to the richer parts of the plains near the city of Parma or to the city proper (and other cities) and neighborhoods, under the stimulus of industrial development and agricultural progress.

The region was subdivided in different ways. Probably the most meaningful demographic unit from a genetic point of view was the commune, which included several small parishes, especially in the most mountainous area. The effect of drift is measured essentially by the variance of gene frequencies of neighboring demes (population units expected to be under random mating for the genes being investigated). Variance errors depend strongly on the number of observations, and the most meaningful observations were obtained by grouping villages in three areas of approximately equal size: plains, hills, and mountains.

Genes studied were those easily determined at the time, belonging to three blood group systems: ABO, MN, and RH. Altogether, 11 different independent allele frequencies were calculated, of which only 8 were of sufficient magnitude to be really useful. A significantly greater value of the observed variance compared with the theoretical one, calculated from the demographic data, was found in the population of the upper Parma valley (the mountainous part). The variances among parishes observed in the plains did not indicate significant heterogeneity, taken to mean that no drift was observed. The hills gave intermediate drift values, but it was advantageous to pool them with the plains.

The mountain data originated from a population of about 5,000 individuals subdivided into 22 villages, the demography of which could be examined for the last 250 years. Calculations were repeated to clarify an earlier incomplete agreement (Cavalli-Sforza and Bodmer 1971, 1999) of observed and expected variances. The most satisfactory expectations, generated by computer simulation, show essentially perfect agreement with observation. In the middle part of the valley, the hills, drift is smaller, as expected; in the lowest part of the valley, the plains, where village sizes were larger than in the highest one by a factor of ten, no detectable genetic heterogeneity of the populations was found. This does not mean that there is no drift in such conditions, but with the numbers of individuals and genes tested it is undetectable.

There was little or no evidence of natural selection, and the conclusion that drift was practically the only detectable cause of genetic variation in the Parma valley seems reasonable, but for reaching significant conclusions it was essential to select areas where drift can be expected to be higher. The number of genes used at the time of this research could hardly have been increased, and, unfortunately, blood samples could not be kept for analysis with genetic markers detected later, because of refrigerator failure.

Later work could be done using surnames. An extended survey in Sardinia (Zei et al. 1986) indicated that these, an equivalent of Y-chromosome markers, could be used with considerable advantage, given the high number of "alleles" they supply. Drift for paternally or maternally transmitted markers is on average four times higher than that of autosomal genes like ABO, MN, and RH. In addition, it is known that drift of patrilinearly transmitted genes may be somewhat higher, on average, than that of matrilineally derived ones (Seielstad et al. 1998), and therefore tends also to slightly overestimate drift observed for autosomal genes. Males marry later and die earlier, are on average more polygamous, and, therefore, tend to have more children on average, as well as greater variation per individual than females. They also migrate less, on average, at least in a greater number of societies.

The theory used with surnames, originally suggested by R. A. Fisher for ecological analysis (the number of species), has been shown to be sufficiently similar (Zei et al. 1986) to that of "infinite number of alleles" (Ewens 1972) that estimates derived from one can be almost equated to those obtained by the other. Both theories suffer, however, from an approximation that cannot yet be assessed accurately, because they assume equal mutation rates between pairs of alleles and are affected in ways not always easily assessed by the unknown variation of individual mutation rates.

We have shown that estimations of drift by standard quantities like $F_{ST}$, based on the variance of gene frequencies or surname frequencies among populations (also called Wahlund variances [Cavalli-Sforza and Bodmer 1971, 1999]), standardized by dividing each variance by $p(1 - p)$, where $p$ is the average gene frequency from which the variance was calculated, agree with those obtained by quantities like $Nm$, used to estimate drift expected in a population because of its size and migratory exchange with neighbors, with which they are related inversely. An approximate relationship is $F_{ST} = 1/(1 + Nm)$ for unilinearly transmitted genetic markers like those of the non-recombining portion of the Y chromosome, as well as surnames. It is $1/(1 + 4Nm)$ for autosomal markers. The equivalent of $Nm$ for surnames is the richness or abundance of surnames, $N_S$, calculated on the basis of Fisher's theory. With surnames one can calculate directly both the $F_{ST}$ variance of the frequency of surnames, and $Nm$; the $F_{ST}$ variance can give an independent estimate of the $Nm$ value from $F_{ST} = 1/(1 + Nm)$. The two estimates of $Nm$, from surnames and genes are independent and usually agree reasonably well. Thus, the analysis of surnames supplies a very interesting way of studying drift in a large population that shares a common linguistic background, and for which, therefore, the origin of surnames is reasonably uniform, as is true of most European countries.

The test of drift with surnames has given us the chance of obtaining two results of general interest. The first was obtained when we compared estimates of drift obtained with surnames with those obtained with genes (chapter 5 and 10). The two independent estimates of $Nm$ obtained with surnames, from $F_{ST}$ and from a direct calculation of $N_S$, agreed between themselves in the test on the Parma Valley, and also agreed with that of our three autosomal genes. The second was that the availability of the surname of 95% Italian families through telephone companies has given us a chance to prepare a geographic map of drift in appropriate units in which we subdivided the Italian population (demes). The natural choice for subdivision was the smallest administrative unit, the commune, of which there exist over 8,000 in Italy. The commune seems a reasonable choice for the demographic unit to be used for genetic purposes, mostly because it is not very far from the desired random breeding unit. There is, moreover, statistical information on each commune for a great number of ecological and socioeconomic variables, which can be valuable for correlation studies.

In the next section we discuss the term and concept of deme, which we consider synonymous with the older term "Mendelian population," and why we use it here to describe the maps of drift by commune.

One can see that the smallest demes are ordinarily located at high altitudes, but tourism has magnified greatly the size estimates of certain demes in some well-known mountain and ski resorts. In the plains and particularly along the coasts, touristic, urban, and industrial development have caused many other demes to swell. Some of this swelling is entirely artificial, being due to second homes and other reasons of amplification of telephone services. Other sources of surname data, for example, voter lists, when available, parish books, and so on, might provide information less easily subject to this kind of disturbance. But these are not, or not regularly, available on computer files, and are therefore useful only for statistics in very limited areas.

The measure of drift was the abundance of surnames per commune ($N_S$), after appropriate corrections that eliminate possible sources of error, for instance, excluding singletons, that is, surnames appearing in only one individual, to decrease the bias generated by recent immigration (Zei et al. 1986). There is a variation of about 1:1,000 between the $N_S$ of the smallest commune and that of the largest. $N_S$ is correlated with the number of inhabitants but tends to be a smaller number. It is likely to be underestimated for those communes that have recently undergone partial depopulation, and overestimated for cities and large towns.

Moreover, our drift maps might be corrected for the erroneous increase of $N_S$ introduced by the inflation of the number of surnames of people who have recently acquired second homes in the area, and other very recent additions to the original population due to tourism or professional reasons. In part, this is achieved by the elimination of singletons. A more satisfactory way is to dissect the distribution of surnames in a commune subject to this disturbance into a fraction of "autochtonous" surnames and a residual, made of recent immigrants. We have recently carried out an analysis for recognizing, for other genetic purposes, the surnames of "autochtonous" individuals, who have been in an area for a few centuries, ideally since the local origin of surnames. The method is based on the selection of individuals whose surname falls into a distribution range varying between, say 50 and 5,000 individuals in the whole country. These are, at least in Italy, a majority of all surnames, and the analysis of their geographic distribution by commune shows that a good fraction, not far from 50%, are concentrated in a narrow geographic area, very likely to be close to their place of origin, or at least of first immigration to the area. The reason is that, especially in older times, change of residence was infrequent, and most marriages were between individuals born in neighboring places, often the same village. A fraction of the surname carriers migrated at an earlier time to towns and cities, often the nearest ones, but also the few bigger ones offering more opportunities of jobs and entertainment. Thus, the analysis of the geographic distribution of each surname allows us to concentrate on surnames that are more informative for a variety of genetic purposes. It is, of course, necessary to avoid the introduction of biases due to the selection of surnames. We have not yet carried out the necessary work on a sufficient scale, and at the moment we must accept the fact that areas of the geographic map of Italy that have received a recent flow of part- or full-time residents because of tourism and professions linked with it, or other reasons of immigration, such as industrial or other type of economic development, have $N_S$ values larger than they deserve. These places are, however, well known and easily identified. In general, one can conclude that all the high $N_S$ values are the most unsatisfactory ones, because they betray a considerable heterogeneity of the local population and therefore a deviation from random mating.

Analysis of the results shows considerable variation of deme size, which is profoundly affected by ecological variation (epitomized by altitude), and by all factors that influence population size. Among these, tourism and industrial development have been a major force in Italy. If we had data allowing a similar analysis 50 or 100 years ago, the geography of demes would have been profoundly different. The work we did using telephone books of year 1993 available on computer tapes could be done for the last three centuries using parish books. It would be much more informative if one were to use the lowest unit of geographic clustering, the parish. It would have to be streamlined considerably, and its size reduced by taking samples for periods if not for areas, to make it reasonably efficient. In England, much demographic analysis based on parish books was done by volunteers. Here is a task for an ambitious and dedicated organizer.

Information on drift can be useful for the study of genetic diseases, since demes with small $N_S$ might be isolates lending themselves particularly well to the analysis of complex genetic disorders unusually frequent in them.

## 11.4 DEMES, ISOLATES, AND MIGRATION

The word deme is not widely used, and has never received a standard, practical definition. It is nice and short. It is usually conceived as synonymous with a Mendelian, or random breeding, population. Randomness is usually tested by validity of the Hardy-Weinberg (HW) expectation. One generation of random mating is sufficient for generating HW expectation for single genes, even starting from a heterogeneous population, but randomness can be expected only for invisible genes, since there is almost always some assortative mating for visible traits, indicating internal heterogeneity. Moreover, if random mating could be tested very accurately, some heterogeneity would be expected in almost every population of nontrivial size, because genetic distance always increases with geographic distance (Cavalli-Sforza et al. 1994, section 2.9). By contrast, the great majority of samples tested for classical genes are found to be under random mating by HW tests, except for very large samples such as are collected usually only for blood donors from large areas, such as an entire country. This genetic heterogeneity is almost inevitable because of the greater likelihood of finding significant heterogeneity when the samples are large, and also because of the geographic variation when samples cover a large geographic area.

The main use for tests of random mating by HW is for finding evidence of strong selection, strong heterogeneity, or inbreeding (but further tests are necessary for distinguishing the two, since both cause a relative loss of heterozygotes), or for the validity of hypotheses on the genetic system when many alleles are present.

The main interest in studying certain small, often isolated populations is to evaluate their susceptibility to drift and therefore to estimate their "isolate size," basically a demographic question. Human populations are particularly useful for the study of drift, because head counting is a frequent and superficially simple exercise. Census size is one of the two basic ingredients for estimating drift, after reducing it to effective population size. The other is migration, which reduces drift. It is very rare that an isolate is truly entirely isolated, and some in-migration may be well hidden, although genetic tests might reveal it. To evaluate expected drift not just a census size, but the frequency of in-migrants per generation is also necessary, and the useful parameters for measuring drift (on an inverse scale) are $Nm$ and/or $N_S$. We keep them separate, although they are expected to be very similar, because $N_S$ refers only to males and both $N$ and $m$ are somewhat different for males and females. $Nm$ should be a little smaller, on average, for males.

One observation we have repeatedly made in the course of this book is that migration is far from constant, but is correlated negatively with the size of the population unit, and therefore also with its effective population size $N$ (see also Wijsman and Cavalli-Sforza 1984). This is very reasonable. Inhabitants of small villages frequently tend to look outside the village for spouses. Some villages are so small that the choice of a spouse, that is, an unmarried person, who is not too far from one's age and has other desirable requisites may be limited, inside the village, to very few individuals, say between zero and five. As the size of the population unit increases, or people have opportunities to look outside the bondaries of their village, there are more chances of finding the right person. We have seen that it is only major cities that have a greater in-migration than would be expected given their size, because they usually provide greater opportunities for jobs and entertainment, the two major drawing forces of cities, and therefore tend to absorb immigrants from a wider area. They will therefore tend to be outliers in the correlation.

The negative correlation between $N$ and $m$ decreases the rate of increase of the product $Nm$ with the value of $N$, making it less than simply proportional to $N$. It also helps solving a problem of choosing a level of population cluster for the analysis. In fact, there is always a problem of choosing the best clustering level in a fairly large scale of possible ones. We have used various levels of population units: communes, provinces, and regions in the civil hierarchy, and parishes and dioceses in the religious ones. Unfortunately, the two scales cannot be made compatible with each other except by quantitative compromise, or else we could have an excellent extended scale. It is to a large extent arbitrary which level of clustering is chosen for the deme, but, naturally, the lower the level, the closer the population unit is to random mating. This, however, is not so important because there would be some degree of internal heterogeneity at every level. Provided we have intermigration estimates (ideally as full migration matrices and at worst as average estimate of migration from one unit to an adjacent one) we can obtain estimates of $Nm$. This is the quantity desired as estimates of drift from gene frequencies or other values.

There is thus a range of possible choices of the unit to be used as a deme, and if we have estimates from more than one level we have additional ways of estimating, for instance, internal heterogeneity, variance of gene frequencies and their derivatives, inbreeding coefficients, and other quantities that help us to understand and measure important aspects of genetic population structure. It seems, therefore, that the choice of a deme is potentially wide and to some extent arbitrary, and ideally one is interested in choosing it at the clustering level at which both census size and migration estimates are available, to evaluate $N$, $m$, their correlation, and $Nm$ values or equivalent ones.

Of the three major evolutionary forces—mutation, natural selection, and drift (migration, often named separately from drift as the fourth force, is really included in it by estimating $Nm$ values)—mutation is the one of which we know more quantitatively (at least about orders of magnitude of frequencies). Of natural selection and drift we still have only modest quantitative knowledge. The distribution of Darwinian fitness values of mutants may help to predict the relative evolutionary effects of drift and selection. A major breakthrough was Kimura's observation (1968, 1983) that, in many cases, estimates of protein evolutionary rates are not very far from those predicted on the basis of mutation rates, a result expected if most protein evolution were due to mutation and drift alone. This gave further support to the idea that the fraction of mutations with small or zero effect on Darwinian fitness cannot be far from trivial. The best scale on which to compare natural selection and drift is perhaps that of the variance of gene frequencies, and it would seem that drift gives a nontrivial contribution to the acceptance of polymorphic mutations (Cavalli-Sforza et al. 1994).

In spite of the variety of methods proposed for the demonstration of natural selection based on the analysis of DNA polymorphisms, it has so far been almost impossible to show conclusively that common human polymorphisms are the result of selective advantage of new mutations, except for classical mutants whose selective advantage due to resistance to infectious diseases was directly proved earlier (Sabeti et al. 2002). It is reasonable to expect, however, that studies of genomics/proteomics will help to throw more light on the relative roles of natural selection and drift in generating the variation observed. We hope that our observations and methods may help to expand the understanding of the quantitative contribution, of drift and selection to human variation.

# Bibliography

Adams, M. S., and J. V. Neel. 1967. Children of incest. *Pediatrics* 40(1): 55–62.

Angioni, D., S. Loi, G. Puggioni. 1997. *La popolazione dei comuni sardi dal 1688 al 1991*. Cagliari: CUEC.

Aricò, M., O. Fiorani, A. Lisa, V. Spica Russotto, S. Varotto, V. Conter, C. De Fusco, P. D. D'Angelo, G. Zei, and C. Danesino. 2000. Incidence and geographic distribution of HLH in Italy. *Med. Pediatr. Oncol.* 36:419.

Arner, G. B. 1908. Consanguineous marriages in the American population. *Columbia Univ. Stud. Hist. Econ. Public Law* 31(3): 1–99.

Bahuchet, S. 1979. Utilisation de l'espace forestier par les Pygmées Aka, Chasseur-cueilleurs d'Afrique centrale. *Inform. Sci. Sociales* 18: 999–1019.

Barrai, I., and A. Moroni. 1965. Variazione secolare della consanguineità nella diocesi di Reggio Emilia. *Atti Assoc. Genet. It.* 10: 320–326.

Barrai, I., L. L. Cavalli-Sforza, and A. Moroni. 1962. Frequencies of pedigrees of consanguineous marriages and mating structure of the population. *Ann. Hum. Genet.* 25: 347–376.

Barrai, I., L. L. Cavalli-Sforza, and M. Mainardi. 1964. Testing a model of dominant inheritance for metric traits in man. *Heredity* 19(4): 651–668.

Barrai, I., M. P. Mi, N. E. Morton, and N. Yasuda. 1965. Estimation of prevalence under incomplete selection. *Am. J. Hum. Genet.* 17: 221–236.

Beadle, G. W., and E. L. Tatum. 1941. Genetic control of biochemical reactions in neurospora. *Proc. Natl. Acad. Sci. USA* 27: 499–506.

Bemiss, S. M. 1858. Report on influence of marriages of consanguinity upon offspring. *Trans. Am. Med. Assoc.* 11: 319–425.

Bigozzi, U., C. Conti, R. Guazzelli, E. Montali, and F. Salti. 1970. Morbilità e mortalità nella prole di 300 coppie di coniugi consanguinei nel comune di Firenze. *Acta Genet. Med. Gemellol.* 19: 515–528.

Bittles, A. H. 1994. The role and significance of consanguinity as a demographic variable. *Popul. Dev. Rev.* 20: 561–584.

Bittles, A. H. 2001. Consanguinity and its relevance to clinical genetics. *Clin. Genet.* 60: 89–98.

Bittles, A. H., and J. V. Neel. 1994. The costs of human inbreeding and their implications for variation at the DNA level. *Nat. Genet.* 8: 117–121.

Bodmer, W. F., and L. L. Cavalli-Sforza. 1968. A migration matrix model for the study of random genetic drift. *Genetics* 59: 565–592.

Bodmer, W. F., and L. L. Cavalli-Sforza. 1976. *Genetics, evolution, and man.* San Francisco: W. H. Freeman.

Boiardi, G. 1961. Aspetti sociali dei matrimoni tra consanguinei. Thesis, University of Parma, Italy.

Bonfante, P. 1925. *Corso di diritto romano 1: Diritto di famiglia.* Roma: A. Sampaolesi.

Bonné-Tamir, B. 1980. The Samaritans—a living ancient isolate. In: *Population structure and genetic disorders.* A. W. Eriksson et al., eds. London: Academic Press, 27–41.

Braglia, G. L. 1962. *Frequenze di matrimoni consanguinei.* Thesis, University of Parma, Italy.

Brass, W. 1958. The distribution of births in human populations. *Pop. Stud.* 12:51–72.

Bruce-Chwatt, L. J., and J. De Zulueta. 1980. *The rise and fall of malaria in Europe.* Oxford, UK: Oxford University Press.

Brunet, G., P. Darlu, and G. Zei, eds. 2001. *Le patronyme. Histoire, anthropologie, société.* Paris: CNRS Editions.

Burrows, M. 1938. *The basis of Israelite marriage.* New Haven, CT: American Oriental Society.

Cantoni, G. 1931. Su la consanguineità nelle valli alpestri della Venezia Tridentina. *Atti del Congresso Intern. per gli studi sulla Popolazione,* Roma, estratto p. 7.

Cantoni, G. 1935. Ricerche sulla consanguineità in valli alpestri della Venezia Tridentina. *Genus* 1:251–358.

Cantoni, G. 1936. Su di un paese altamente consanguineo dell'Alta Val Venosta. *Boll. Soc. It. Biol. Sperim.* 11: 284–286.

Cantoni, G. 1938. Ricerche su di un piccolo aggregato umano altamente consanguineo. *Stud. Trentini Sci. Nat.* 19: 1–37.

Cavalli-Sforza, L. L. 1957. Some notes on breeding patterns of human population. *Acta Genet.* Basel 6: 395–399.

Cavalli-Sforza, L. L. 1958. Some data on genetic structure of human populations. *Proc. Tenth. Int. Congr. Genet.* 1: 389–407.

Cavalli-Sforza, L. L. 1960. Indagine speciale sulla consanguineità dei matrimoni. Note e relazioni, n. 11. Istituto Centrale di Statistica, Roma.

Cavalli-Sforza, L. L. 1962. The distribution of migration distances, models and applications to genetics. In: *Human displacements; measurement, methodological aspects.* J. Sutter, ed., pp. 139–158. Entretiens de Monaco en Sciences Humaines. Monaco: Editions Sciences Humaines.

Cavalli-Sforza, L. L. 1966. Population structure and human evolution. *Proc. R. Soc. London B* 164: 362–379.

Cavalli-Sforza, L. L. 1967. Human populations. In: *Heritage from Mendel.* A. Brink, ed. Madison: University of Wisconsin Press.

Cavalli-Sforza, L. L. 1969. Genetic drift in an Italian population. *Sci. Am.* 221(2): 30–37.

Cavalli-Sforza, L. L. 1986. *African Pygmies*. Orlando, FL: Academic Press.

Cavalli-Sforza, L. L. 2001. Pourquoi les patronymes? In: *Le Patronyme—Histoire, anthropologie, société*. G. Brunet, P. Darlu, and G. Zei, eds. pp. 407–418. Paris: CNRS Editions.

Cavalli-Sforza, L. L., and W. F. Bodmer. 1971, 1999. *The genetics of human populations*. San Francisco: W. H. Freeman.

Cavalli-Sforza, L. L., and M. W. Feldman. 1990. Spatial subdivision of populations and estimates of genetic variation. *Theor. Biol.* 37: 3–25.

Cavalli-Sforza, L. L., and B. Hewlett. 1982. Exploration and mating range in African Pygmies. *Ann. Hum. Genet.* 46: 257–270.

Cavalli-Sforza, L. L., and G. Zei. 1967. Experiments with an artificial population. *Proc. Third. Int. Congr. Hum. Genet.* pp. 473–478. Baltimore, MD: The Johns Hopkins Press.

Cavalli-Sforza, L. L., M. Kimura, and I. Barrai. 1966. The probability of consanguineous marriages. *Genetics* 54: 37–60.

Cavalli-Sforza, L. L., P. Menozzi, and A. Piazza. 1994. *History and geography of human genes*. Princeton, NJ: Princeton University Press.

Cerimele, D., F. Cottoni, S. Scappaticci, G. Rabbiosi, G. Borroni, E. Sauna, G. Zei, and M. Fraccaro. 1982. High prevalence of Werner's syndrome in Sardinia. Description of six patients and estimate of the gene frequency. *Hum. Genet.* 62: 25–30.

Chen, K. H., and L. L. Cavalli-Sforza. 1983. Surnames in Taiwan: interpretation based on geography and history. *Hum. Biol. Oceania* 55: 367–374.

Chessa, L., A. Lisa, O. Fiorani, and G. Zei. 1994. Ataxia-telangiectasia in Italy: genetic analysis. *Int. J. Radiat. Biol.* 66(6): 1–33.

Chung, C. S., O. W. Robison, and N. E. Morton. 1959. A note of deaf mutism. *Ann. Hum. Genet.* 23: 357–366.

Cipolla, C. M. 1952–1953. The decline of Italy: the case of a fully matured economy. *Econ. Hist. Rev.* 5: 178–187.

Coale, A. J., and S. Cotts Watkins. 1986. *The decline of fertility in Europe*. Princeton, NJ: Princeton University Press.

Conterio, F. 1967. Effetti della consanguineità: Studi su alcune popolazioni italiane. *Atti Assoc. Genet. It.* 12: 223–235.

Conterio, F., and I. Barrai. 1966. Effetti della consanguineità sulla mortalità e sulla morbilità nella popolazione della diocesi di Parma. *Atti Assoc. Genet. It.* 11: 378–391.

Conterio, F., and L. L. Cavalli-Sforza. 1960. Selezione per caratteri quantitativi nell'uomo. *Atti Assoc. Genet. It.* 5: 295–304.

Conterio, F., and G. Zei. 1964. Consanguineità e malattie mentali. *Atti Assoc. Genet. It.* 9: 224–232.

Crow, J. F. 1958. Some possibilities for measuring selection intensities in man. *Hum. Biol.* 30: 1–13.

Crow, J. F., and A. P. Mange. 1965. Measurement of inbreeding from the frequency of marriages between persons of the same surname. *Eugen. Quart.* 12: 199–203.

Crumpacker, D. W., G. Zei, A. Moroni, and L. L. Cavalli-Sforza. 1976. Air distance versus road distance as a geographical measure for studies on human population structure. *Geogr. Anal.* 8: 215–223.

Dahlberg, G. 1926. Inbreeding in man. *Genetics* 14: 421–454.

Dahlberg, G. 1948. *Mathematical methods for population genetics*, Basel: S. Karger.

Damon, A. 1969. Race, ethnic groups and disease. *Soc. Biol.* 16(2): 69–80.

Danubio, M. E., A. Piro, and A. Tagarelli. 1999. Endogamy and inbreeding since the 17th century in past malarial communities in the province of Cosenza (Calabria, Southern Italy). *Ann. Hum. Biol.* 26(5): 473–488.

De Benedetto, G., I. S. Nasidze, M. Sterico, L. Nigro, M. Krings, M. Lanzinger, L. Vigilant, M. StoneKing, S. Paabo, and G. Barbujani. 2000. Mitochondrial DNA sequences in prehistoric human remains from the Alps. *Eur. J. Hum. Genet.* 8(9): 669–677.

Dobzhansky, T. 1951. *Genetics and the origin of species.* New York: Columbia University Press.

Du, R., Y. Yuan, J. Hwang, J. Mountain, and L. L. Cavalli-Sforza. 1992. Chinese surnames and their genetic differences between North and South China. *Journal of Chinese Linguistics*, Monograph Series 5.

Edwards, A. W. F. 1968. Simulation studies of genealogies. *Heredity* 23: 628.

Esmein, A. 1933. *Le marriage en droit canonique.* Paris: Recueil Sirey.

Ewens, W. J. 1972. The sampling theory of selectively neutral alleles. *Theor. Popul. Biol.* 3: 87–112.

Fenoglio, S. 1956. Consanguineità e sterilità. *Minerva Med.* 56: 229–232.

Fenoglio, S. 1969. Consanguineità e gruppi sanguigni. *Minerva Med.* 60(84): 4190–4194.

Fisher, R. A. 1943. The relation between the number of species and the number of individuals in a random sample of an animal population, part 3. *J. Anim. Ecol.* 12: 42–58.

Fleury, J. 1933. *Recherches historiques sur les empechements de parente dans le mariage canonique des origines aux Fausses decretales.* Paris: Recueil Sirey.

Fraccaro, M. 1957. Consanguineous marriages in Italy: a note. *Eugen. Quart.* 4: 36–39.

Francalacci, P., L. Morelli, P. A. Underhill, A. S. Lillie, G. Passarino, A. Useli, R. Madeddu, G. Paoli, S. Tofanelli, C. M. Calò, M. E. Ghiani, L. Varesi, M. Memmi, G. Vona, A. A. Lin, P. Oefner, and L. L. Cavalli-Sforza. 2003. Peopling of three Mediterranean islands (Corsica, Sardinia, and Sicily) inferred by Y-chromosome biallelic variability. *Am. J. Phys. Anthropol.* 121: 270–279.

Frota-Pessoa, O. 1957. The estimation of the size of isolates based on census data. *Am. J. Hum. Genet.* 2: 9–16.

Garrod, A. E. 1902. The incidence of alkaptonuria: a study in chemical individuality. Reprinted in S. H. Boyer, ed. *Papers on human genetics*. Englewood Cliffs, NJ: Prentice Hall.

Gianferrari, L. 1932. Sul metodo proposto dal Krizenecky applicato al calcolo del coefficiente di consanguineità della popolazione italiana entro i vecchi confini del regno. *Atti Soc. It. Sci. Nat.* 71: 8.

Gianferrari, L. 1936. Sugli effetti demografici della consanguineità in valle Venosta e nelle valli laterali. *Atti Soc. It. Sci. Nat.* 75: 33–43.

Gibbons, A. 1993. The risks of inbreeding. *Science* 259: 1252.

Gilmour, J. S. L., and J. W. Gregor. 1939. Demes: a suggested new terminology. *Nature* 144: 333.

Gottlieb, K., ed. 1983. Surnames as markers of inbreeding and migration. *Hum. Biol.* 55(2).

Grant, J. C., and A. H. Bittles. 1997. The comparative role of consanguinity in infant and childhood mortality in Pakistan. *Ann. Hum. Genet.* 61: 143–149.

Guerresi, P., D. Pettener, and F. M. Veronesi. 2001. Marriage behaviour in the Alpine Non Valley from 1825 to 1923. *Ann. Hum Biol.* 28(2): 157–171.

Guglielmino Matessi, C., and G. Zei. 1979. Un metodo di correzione per il calcolo delle frequenze geniche in piccole popolazioni composte da nuclei familiari. *Atti Assoc. Genet. It.* 24.

Hajnal, J. 1963. Random mating and the frequency of consanguineous marriages. *Proc. R. Soc. London B* 159: 125–177.

Jorde, L. B. 1980. The genetic structure of subdivided human populations: a review. In: *Current developments in anthropological genetics*, Vol. 1. J. H. Mielke and M. H. Crawford, eds., pp. 135–208. New York: Plenum.

Joyce, G. H. 1948. *Christian marriage: an historical and doctrinal study*. London: Sheed & Ward.

Karlin, S. 1966. *A first course in stochastic processes*. New York: Academic Press.

Karlin, S., and J. McGregor. 1967. The number of mutant forms maintained in a population. *Proc. Fifth Berkely Symp. Math. Stat. Prob.* 4: 415–438.

Keyfitz, N. 1985. *Applied mathematical demography*, 2nd ed. New York: Springer-Verlag.

Kimura, M. 1968. Evolutionary rate at the molecular level. *Nature* 217: 624–626.

Kimura, M. 1983. *The neutral theory of molecular evolution*. Cambridge, UK: Cambridge University Press.

Lasker, G. W. 1985. *Surnames and genetic structures*. Cambridge, UK: Cambridge University Press.

Laughlin, W. S. 1950. Blood groups, morphology and population size of the Eskimos. *Cold Spring Harbor Symp. Quant. Biol.* 15: 165–173.

Leone, M., F. Brignolio, M. G. Rosso, E. S. Curtoni, A. Moroni, A. Tribolo, and D. Schiffer. 1990. Friedreich's ataxia: a descriptive epidemiological study in an Italian population. *Clin. Gen.* 38: 161–169.

Lilliu, G. 1983. *La civiltà dei Sardi: Dal Neolitico all'età dei Nuraghi.* Torino: ERI.

Lisa, A., O. Fiorani, and G. Zei. 1996. Surname distributions give an estimate of the degree of genetic population structure. *Braz. J. Genet.* 19(2): 116.

Lisa, A., O. Fiorani, E. Siri, and G. Zei. 2001. Migrations récentes en Italie et distribution géographique des patronymes. In: *Le patronyme: histoire, anthropologie, société.* G. Brunet, P. Darlu, G. Zei, eds., pp. 231–244. Paris: CNRS Editions.

Livi, R. 1896. *Antropometria militare.* Roma, Giornale del Regio Esercito.

Livi-Bacci, M. 1977. *A history of Italian fertility.* Princeton, NJ: Princeton University Press.

Mainardi, M., L. L. Cavalli-Sforza, and I. Barrai. 1962. The distribution of the number of collateral relatives. *Atti Assoc. Genet. It.* 7: 123–130.

Malécot, G. 1950. Quelques schémas probabilistes sur la variabilité des populations naturelles. *Ann. Univ. Lyon Sci. A.* 13: 37–60.

Mange, A. P. 1964. Growth and inbreeding of a human isolate. *Hum. Biol.* 36: 104–133.

Mantegazza, P. 1868. *Studj sui matrimonj consanguinei.* Milano: G. Brignola.

Moroni, A. 1967. Struttura ed evoluzione della consanguineità umana nelle isole Eolie (1680–1966). *Archiv. Antropol. Etnol.* 98(3): 135–150.

Moroni, A., and P. Menozzi. 1972. La consanguineità umana in Sicilia. *Ateneo Parmense* 8(1): 3–39.

Moroni, A., A. Anelli, W. Anghinetti, E. Lucchetti, O. Rossi, and E. Siri. 1972. La consanguineità umana nell'isola di Sardegna dal secolo XVIII al secolo XX. *Ateneo Parmense* 8(1): 69–92.

Moroni, A., A. Anelli, W. Anghinetti, and E. Siri. 1992. *I matrimoni tra consanguinei.* Napoli: Ed. Comunicazioni Sociali.

Morton, N. E. 1955. Non-randomness in consanguineous marriage. *Ann. Hum. Genet.* 20: 116–124.

Morton, N. E., J. F. Crow, and H. J. Muller. 1956. An estimate of the mutational damage in man from data on consanguineous marriages. *Proc. Natl. Acad. Sci USA* 42: 855–863.

Morton, N. E., R. Lew, I. E. Husserl, and G. F. Little. 1972. Pingelap and Mokil Atolls: historical genetics. *Am. J. Hum. Genet.* 24: 277–289.

Mourant, A. E., A. C. Kopec, and K. Domaniewska-Sobczac. 1958. *The ABO blood groups.* Oxford, UK: Blackwell.

Mourant, A. E., A. C. Kopec, and K. Domaniewska-Sobczac. 1976. *The distribution of the human blood groups and other polymorphisms.* London: Oxford University Press.

Muller, H. J. 1950. Our load of mutation. *Am. J. Hum. Genet.* 2(2): 111–176.

Murdock, G. P. 1967. *Ethnographic atlas*. Pittsburgh, PA: University of Pittsburgh Press.

Murru Corriga, G. 1990. *Dalla montagna ai Campidani*. Cagliari: EDES.

Nei, M., and J. Imaizumi. 1966. Genetic structure of human populations. *Heredity* 21: 183–190.

Neufeld, E. 1944. *Ancient Hebrew marriage laws, with special references to general Semitic laws and customs*. London: Longmans, Green.

Paglino, G. 1952. *La famiglia presso gli Ebrei e altri popoli semitici*. Alba, Italy: Edizioni Paoline.

Penot, 1902. *Evolution du mariage et consanguinité*. Lyon: A. Storck.

Penrose, L. S. 1963. *The biology of mental defects*, 3rd ed. New York: Grune & Stratton.

Pettener, D. 1985. Consanguineous marriages in the Upper Bologna Appennine (1565–1980): microgeographic variations, pedigree structure and correlation of inbreeding secular trend with changes in population size. *Hum. Biol.* 57: 267–288.

Piazza, A., S. Rendine, G. Zei, A. Moroni, and L. L. Cavalli-Sforza. 1987. Migration rates of human populations from surname distribution. *Nature* 329: 714–716.

Piazza, A., N. Cappello, E. Olivetti, and S. Rendine. 1988. A genetic history of Italy. *Ann. Hum. Genet.* 52: 203–213.

Reynolds, J., B. S. Weir, and C. C. Cockerham. 1983. Estimation of the coancestry coefficient: basis for a short-term genetic distance. *Genetics* 47: 329–352.

Roberts, D. F. 1968. Genetic effects of population size reduction. *Nature* 220: 1084–1088.

Romeo, G., P. Menozzi, A. Ferlini, S. Fadda, S. Di Donato, G. Uziel, B. Lucci, L. Capodaglio, A. Filla, and G. Campanella. 1983a. Incidence of Friedreich Ataxia in Italy estimated from consanguineous marriages. *Am. J. Hum. Genet.* 35: 523–529.

Romeo, G., P. Menozzi, A. Ferlini, L. Prosperi, R. Cerone, S. Scalisi, C. Romano, I. Antonozzi, E. Riva, L. Piceni Sereni, E. Zammarchi, G. Lenzi, R. Sartorio, G. Andria, M. Cioni, A. Fois, M. Burroni, A. B. Burlina, and F. Carnevale. 1983b. Incidence of classic PKU in Italy estimated from consanguineous marriages and from neonatal screening. *Clin. Genet.* 24: 339–345.

Romeo, G., M. Bianco, M. Devoto, P. Menozzi, G. Mastella, A. M. Giunta, C. Micalizzi, M. Antonelli, A. Battistini, F. Santamaria, D. Castello, A. Marianelli, A.G. Marchi, A. Manca, and A. Miano. 1985. Incidence in Italy, genetic heterogeneity, and segretation analysis of cystic fibrosis. *Am. J. Hum. Genet.* 37: 338–349.

Sabeti, P. C., D. E. Reich, J. M. Higgins, H. Z. Levine, D. J. Richter, S. F. Schaffner, S. B. Gabriel, J. V. Platko, N. J. Patterson, G. J. MacDonald, H. C. Ackerman, S. J. Campbell, D. Altshuler, R. Cooper, D. Kwiatkowski, R. Ward, and E. S. Lander. 2002. Detecting recent positive selection in the human genome from haplotype structure. *Nature* 419(6909): 832–837.

Sanchez, T. 1607. *De Sancto Matrimonii Sacramento.* Antverpiae.

Schull, W. J., and J. V. Neel. 1965. *The effect of inbreeding on Japanese children.* New York: Harper & Row.

Seielstad, M. T., E. Minch, and L. L. Cavalli-Sforza. 1998. Genetic evidence for higher female migration rate in humans. *Nat. Genet.* 20(3): 278–280.

Serra, A. 1959. Letter to Editor: Italian official statistics on consanguineous marriages. *Acta Genet.* 9: 244–246.

Serra, A., and A. Soini. 1959. La consanguinité d'une population. Rappel de notions et de résultats. Application à trois provinces de l'Italie du Nord. *Population* 14: 47–72.

Serra, A., and A. Soini. 1961. La consanguineità nel Lodigiano dal 1900 al 1956. *A. Ge. Me. Ge.* 10(4): 485–501.

Siniscalco, M., L. Bernini, L. B. Latte and A. G. Motulsky. 1961. Favism and thalassemia in Sardinia and their relationship to malaria. *Nature* 190: 1179–1180.

Siniscalco, M., R. Robledo, P. K. Bender, C. Carcassi, L. Contu, and J. C. Beck. 1999. Population genomics in Sardinia: a novel approach to hunt for genomic combinations underlying complex traits and diseases. *Cytogenet. Cell Genet.* 86: 148–152.

Skolnick, M. H. 1974. The construction and analysis of genealogies from parish registers with a case study of Parma valley, Italy. Thesis, Stanford University, Palo Alto, CA.

Skolnick, M. H., A. Moroni, C. Cannings and L. L. Cavalli-Sforza. 1971. The reconstruction of genealogies from parish books. In: *Mathematics in the archaeological and historical sciences.* F. R. Hodson et al., eds., pp. 319–334. Edinburgh: Edinburgh University Press.

Spoor, C. F., and P. Y. Sondaar. 1986. Human fossil from the endemic island fauna of Sardinia. *J. Hum. Evol.* 15: 399–408.

Sutter, J., and L. Tabah. 1956. Méthode mécanographique pour établir la généalogie d'une population. *Population* 3: 515–520.

Underhill, P. A., L. Jin, A. A. Lin, S. Q. Mehdi, T. Jenkins, D. Vollrath, R. W. Davis, L. L. Cavalli-Sforza, and P. J. Oefner. 1997. Detection of numerous Y chromosome biallelic polymorphisms by denaturing high-performance liquid chromatography. *Genome Res.* 7: 996–1005.

Waddingens, L. 1891. *Johannis Duns Scoti Opera Omnia.* Paris: Vives.

Weir, B. S., and C. C. Cockerham. 1984. Estimating $F$-statistics for the analysis of population structure. *Evolution* 38: 1358–1370.

Wijsman, E. M., and L. L. Cavalli-Sforza. 1984. Migration and genetic population structure with special reference to humans. *Annu. Rev. Ecol. Syst.* 15: 279–301.

Wijsman, E. M., G. Zei, A. Moroni, and L. L. Cavalli-Sforza. 1984. Surnames in Sardinia, II: computation of migration matrices from surname distributions in different periods. *Ann. Hum. Genet.* 48: 65–78.

Wobst, H. M. 1974. Boundary conditions for paleolithic social systems: a simulation approach. *Am. Antiquity* 39: 147–178.

Wolf, A. P. 1980. *Marriage and adoption in China, 1854–1945.* Stanford, CA: Stanford University Press.

Wright, S. 1922. Coefficients of inbreeding and relationship. *Am. Natur.* 56: 330–338.

Yasuda, N., and T. Furusho. 1971. Random and nonrandom inbreeding revealed from isonymy study: small cities in Japan. *Am. J. Hum. Genet.* 23: 303–316.

Yasuda, N., L. L. Cavalli-Sforza, M. Skolnick, and A. Moroni. 1974. The evolution of surnames: an analysis of their distribution and extinction. *Theoret. Popul. Biol.* 5: 123–142.

Zei, G., and E. Zanardi. 1986. Migration and genetic variation. In: *African Pygmies*, L. L. Cavalli-Sforza, ed. pp. 339–346. Orlando, FL: Academic Press.

Zei, G., A. Moroni, and L. L. Cavalli-Sforza. 1971. Age of consanguineous marriages. In: *Génétique et populations: hommage a Jean Sutter.* Cahier n. 60. pp. 147–153. Presses Universitaires de France.

Zei, G., C. R. Guglielmino, E. Siri, A. Moroni, and L. L. Cavalli-Sforza. 1983a. Surnames as neutral alleles: observation in Sardinia. *Hum. Biol.* 55(2):357–365.

Zei, G., R. Guglielmino Matessi, E. Siri, A. Moroni, and L. L. Cavalli-Sforza. 1983b. Surnames in Sardinia, I: fit of frequency distribution for neutral alleles and genetic population structure. *Ann. Hum. Genet.* 47: 329–352.

Zei, G, A. Piazza, A. Moroni, and L. L. Cavalli-Sforza. 1986. III: The spatial distribution of surnames for testing neutrality of genes. *Ann. Hum. Genet.* 50: 169–180.

Zei, G., G. Barbujani, A. Lisa, O. Fiorani, P. Menozzi, E. Siri, and L. L. Cavalli-Sforza. 1993. Barriers to gene flow estimated by surname distribution in Italy. *Ann. Hum. Genet.* 57: 123–140.

Zei, G., A. Lisa, O. Fiorani, and C. R. Guglielmino. 1996. Abundance of different species and surname distributions: from statistical model to genetic information. *It. J. Appl. Stat.* 8: 283–292.

# Index

# MONOGRAPHS IN POPULATION BIOLOGY

EDITED BY SIMON A. LEVIN AND HENRY S. HORN

Titles available in the series (by monograph number)

www.ingramcontent.com/pod-product-compliance
Ingram Content Group UK Ltd.
Pitfield, Milton Keynes, MK11 3LW, UK
UKHW022304121224
452420UK00012B/620